The Great Astronomical Revolution

It is the stars;
The stars above us, govern our conditions.

William Shakespeare: King Lear

ABOUT THE AUTHOR

PATRICK (CALDWELL) MOORE was born on 4 March 1923, son of Capt. Charles Caldwell-Moore, MC, and Gertrude Moore (née White). Illness interrupted his schooling, but though he passed Common Entrance for Public School he was unable to go. He now admits that at the start of the war he manipulated his medical for the Forces, and also his age, to enter the RAF, aged 17; he became a Navigator in Bomber Command, and was commissioned in 1940.

His interest in astronomy dated from the age of 6, when he read a book about the subject. At the age of 11, he was elected a member of the British Astronomical Association (the youngest member ever to be elected: exactly 50 years later he became President), and published his first paper, about a section of the Moon's surface, at the age of 13 years, in 1936 - with the help of a 3-inch telescope that he had acquired at the age of ten.

After leaving the RAF in 1945 he set up a private observatory at his home, and concentrated upon observations of the Moon. He also began writing, and his first major book, *Guide to the Moon*, was published in 1948 (in its latest edition it is still in print). In 1957 he was invited to present a monthly series on BBC Television, *The Sky at Night*, and has done so every four weeks since then without a break (easily a world record). In 1959, when the Russians sent their first unmanned space-craft round the Moon and obtained images of the hidden side, Moore's maps of the lunar libration regions were used. During the 1950s and 1960s he continued with lunar mapping, and during the Apollo missions also acted as the BBC television commentator. Between 1966 and 1969 he became Director of the Armagh Planetarium, and when this had been firmly established he returned to Sussex, setting up his observatory in Selsey.

He was President of the British Astronomical Association, 1982-4, and is current Vice-President; he has also acted as Director of the BAA Lunar Section and the Mercury and Venus Section. He is a Fellow of the Royal Astronomical Society, and a member of the International Astronomical Union (Editor, 1986). He is an Honorary Member of the Royal Astronomical Society of New Zealand, of the Royal Astronomical Society of Canada, and of the Russian Astronomical Society. He was awarded the Lorimer Gold Medal in 1963, Guido Horn d'Arturo Medal (Bologna) in 1967, the Goodacre Medal of the BAA in 1982, and the Jackson-Gwilt Medal of the RAS in 1985; also the Roberts-Klumpke Medal of the Astronomical Society of the Pacific in 1987. He is a Fellow of Queen Mary and Westfield College, University of London, and has been awarded DScs from the Universities of Birmingham, Lancaster, Hertfordshire, and (pending) Keele.

He was created OBE in 1966, and CBE in 1988. Minor Planet 2602, discovered in America, is named 'Moore' in his honour.

THE GREAT ASTRONOMICAL REVOLUTION

1543–1687 and the Space Age Epilogue

PATRICK MOORE, CBE, DSc, FRAS
Past President of the British Astronomical Association

Albion Publishing
Chichester

First published in 1994 by
ALBION PUBLISHING LIMITED
International Publishers, Coll House, Westergate, Chichester, West Sussex, England
USA: **PAUL & COMPANY PUBLISHERS CONSORTIUM INC.**, P.O. Box 422,
Concord, MA 01742

The extracts from Stillman Drake's translation of Galileo's Dialogue, are reproduced by
kind permission of Doubleday and Company Inc. New York

British Library Cataloguing in Publication Data

Moore, Patrick
The Great Astronomical Revolution: 1543–1687
and the Space Age Epilogue
I. Title

ISBN 1-898563-18-7   (Albion Publishing) Library Edition

Half-title illustration: *The Sun and the moon; woodcut from the* Nürnberg Chronicle, 1493

Printed in Great Britain by Hartnolls, Bodmin, Cornwall

Contents

Foreword

In writing this book, I have done my best to tell the story of the greatest
scientific upheaval of all time. My method has been to take the principal
characters one by one, and then link in their careers with the overall
pattern of events; I begin in the remote past, and then turn to the five
great pioneers of the revolution–Copernicus of Poland, Tycho Brahe
of Denmark, Kepler of Germany, Galileo Galilei of Italy, and Sir Isaac
Newton of England. Inevitably I have had to rely upon material which
has been previously published, and I must make special mention of the
invaluable translation of Galileo's *Dialogue* made by Stillman Drake;
the facsimile copy of Copernicus' *De Revolutionibus*, published by
Messrs. Macmillan to mark the quincentenary of his birth; and various
more general books, notably one by my old friend Willy Ley, who had
an amazing ability in unearthing little-known facts and presenting them
in readable form. I must also express my thanks to Peter Bellew, who
first suggested that I should write this book.

This is not a technical work; it is not intended to be so, but I hope
that it will prove enjoyable to those who read it.

Patrick Moore

Astronomer with astrolabe, woodcut from the Nürnberg Chronicle, *1493*

The Earth in Space

This book is the story of a revolution. It was not quick, and except in its later stages it was not violent; moreover it happened a long time ago, and was virtually complete well before the end of the seventeenth century. And yet it resulted in probably the most radical change in outlook that mankind has ever known. To his dismay, *homo sapiens* found that he was not the lord of the universe; instead, he was confined to a small planet in the Sun's family.

To say that this came as a surprise is putting things very mildly. It seemed inconceivable; and quite apart from the scientific objections, there were the religious arguments. To dethrone the Earth from its proud central position in the cosmos struck a blow at the very heart of theological doctrine, and Church leaders were not in the least inclined to accept it. They reacted strongly, and any scientist who opposed the official teachings had to be prepared for rough treatment. Giordano Bruno was burned at the stake in 1600; and although he had committed many crimes in the eyes of the Inquisition, his case was not helped by the fact that he persisted in his claim that the Earth moves round the Sun instead of vice versâ.

It was this question of the status of the Earth which caused all the trouble. Philosophers of the ancient world were almost unanimous in believing that it must be the most important of all bodies, and the few dissenters, such as Aristarchus of Samos (of whom more anon) collected very few followers. By the Middle Ages, Aristarchus' wild suggestions about a moving Earth had been more or less forgotten, and all universities and schools adopted the geocentric hypothesis without a second thought. ('Geocentric' means 'earth-centred'; 'heliocentric', a word to be used a great deal in the following pages, means 'sun-centred'.) Then, in 1543, an elderly Polish canon named Mikołaj Kopernik published a book in which he put forward the contrary view – and the battle was joined. It

9

raged for over a hundred years, and did not really end until 1687, with the publication of the immortal *Principia* of Isaac Newton. True, the main strife was over well before that; but in telling the story of a revolution we must trace not only its origins, but also its aftermath.

This is what I propose to do. It is a fascinating tale, and there are some colourful characters in it. We have Kopernik, always known to us as Copernicus, who was a quiet scholar and yet could turn himself into a warrior and a statesman when the occasion demanded; next came Tycho Brahe, the eccentric Dane who had a nose made of gold, silver and wax, and whose island observatory contained such unusual features as a prison and a pet dwarf; then there was Johannes Kepler, who had to spend years in saving his mother from being burned as a witch; and Galileo Galilei, as brilliant as he was tactless, and who became the first of all great astronomers to look at the skies through a telescope. Finally there was Sir Isaac Newton, who, in his own words, saw further than other men because he had been able to stand on the shoulders of giants.

Yet even before starting the story it is, I think, a wise move to introduce the universe as we know it to be today. In the Space Age, when flights to the Moon and rockets sent out toward the planets no longer earn a place of honour on the front pages of the daily newspapers, most people have at least a working knowledge of astronomy, but there are also some common misconceptions. I have lost count of the number of people who have asked me whether I can cast horoscopes, and by no means everyone is clear about the distinction between astronomy and astrology. Unless the initial picture is clear-cut, much of what follows will make no sense at all. To compress an outline of the universe into a few pages is a difficult matter, and means that much must be left out, but I will do my best.

Let us start with the Earth, which is a globe almost 8000 miles in diameter, moving round the Sun at an average distance of just under 93 million miles. This seems a long way, and indeed it is; I very much doubt whether anyone can really appreciate even one million miles. But in astronomy a distance of this kind is negligible, and the Sun is one of our very closest neighbours in space. We have to become used to immense distances and immense spans of time; and this, of course, is something that our ancestors could not bring themselves to do.

The Earth is a planet, and is a member of the Sun's family or Solar System. It is not the only planet. Five others are visible with the naked eye; Mercury and Venus are closer to the Sun than we are, while Mars, Jupiter and Saturn are farther away. In relatively modern times three more planets have been discovered beyond Saturn, and have been

Shower of meteors seen as knights fighting with swords. From Prodigiorum ac Ostentorum Chronicon, *Basle, 1557*

named Uranus, Neptune and Pluto, but as they were not found until well after the time of Newton there is no point in saying much about them here. Uranus is just visible without optical aid if you know exactly where to look for it, but Neptune and Pluto are too remote and dim.

Gaul, we are told, was divided into three parts. The planetary system is divided into two; and to give a better idea of it, the diagram should help (Fig 1). The inner group of planets is made up of four comparatively small and solid worlds, of which the Earth is actually the largest, though it is closely rivalled by Venus. Beyond the path or orbit of Mars there is a wide gap, in which move thousands of dwarf worlds known as asteroids; these again were not found until after the end of our revolution (Ceres, first on the list, came to light in 1801) and all of them are so small that they give the impression of being cosmical débris. Then we come to the four giants, beginning with the largest of all, Jupiter, which is big enough to swallow up thirteen hundred bodies the volume of the Earth and still leave room to spare. Finally, on the edge of the main system, comes Pluto, which is a peculiar little planet seemingly in a class of its own.

In the diagram it is obvious that the orbits of at least some of the planets are not completely circular. The Earth, for instance, has a distance-range

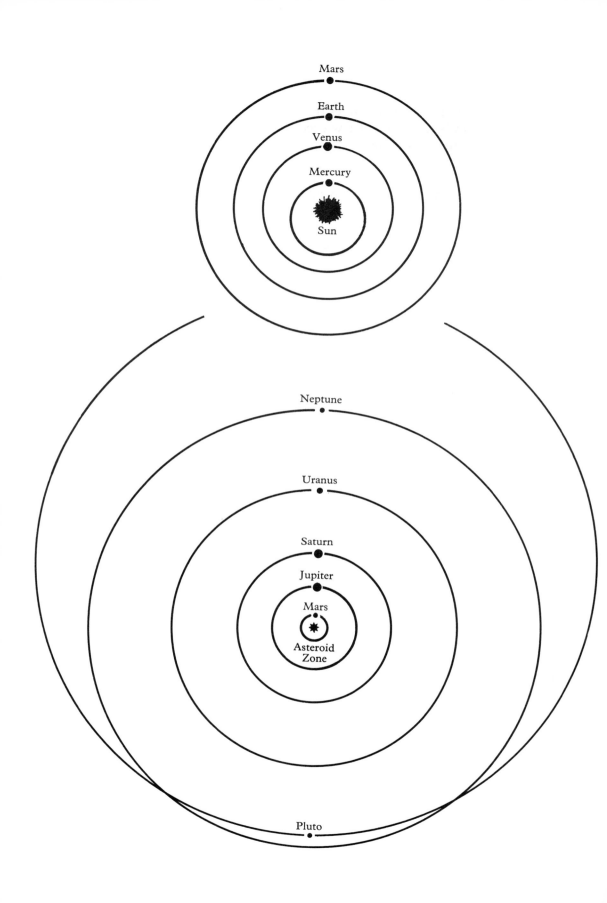

of between $91\frac{1}{2}$ and $94\frac{1}{2}$ million miles from the Sun, so that in technical parlance its orbit is an ellipse of low eccentricity. The difference from the perfectly circular form is not striking, but it is of vital importance, as we will see as the story unfolds.

The planets have no light of their own, and shine only because they reflect the rays of the Sun, so that in the fortunately improbable event of the Sun being suddenly blacked out the planets would vanish too. The same is true of the Moon, which is our faithful companion in space, and stays together with us in our never-ending journey round the Sun. Officially the Moon is ranked as a secondary body or satellite; I have the feeling that the Earth-Moon system would be better regarded as a double planet, but in any case the Moon is a very junior member of the Solar System. It has a diameter of only 2160 miles, and if you could put the Earth into one pan of a gigantic pair of scales you would need eighty-one Moons to balance it.

On average, the Moon's distance from us is a mere 239,000 miles, which is less than ten times the distance round the Earth's equator. This is not very much—and I cannot resist adding that I have an ancient but reliable car whose mileometer registers a grand total of over 600,000, so that I have driven it a distance which is greater than that to the Moon and back. However, the Moon too moves in an orbit which is somewhat elliptical, and so its distance from us is not constant. During one circuit, taking $27\frac{1}{3}$ days, it approaches us to within 230,000 miles and recedes to just over 250,000, or a quarter of a million.

It is an over-simplification to say that the Moon revolves round the Earth. More accurately, the Earth and Moon revolve together round the barycentre, or centre of gravity of the system. The situation may be compared with that of two dumbells being twisted by the bar which joins them. If the bells are equal in weight, the balancing point will be midway between them; if not, then the point will be closer to the heavier bell. With the Earth and Moon, the difference in mass is so great that the barycentre lies inside the Earth's globe, but here again we have a principle whose importance can hardly be over-estimated.

Other plants have satellites of their own; the roll-call is sixteen for Jupiter, eighteen for Saturn, fifteen for Uranus, eight for Neptune, two for Mars and one for Pluto, while Mercury and Venus are solitary travellers. The two attendants of Mars, for instance, are mere ir-regular lumps of material less than twenty miles across, so that they are not even remotely comparable with our Moon. From the point of

Figure 1: Plan of the Solar System.

view of my present theme, the most important of the satellites are the four large moons of Jupiter, which are bright enough to be seen with any telescope; indeed, a good pair of binoculars will show them. They were observed by Galileo in 1610, and played a vital rôle in his defence of the new theories of the universe.

The Solar System also includes the spectacular comets, which are not solid, massive bodies, but are made up of small particles together with very thin gas. It has been said that a comet is the nearest approach to nothing which can still be anything, which is by no means a bad description. There are many small, faint comets which travel round the Sun in eccentric paths and come back into view every few years, but the larger ones have revolution periods of many centuries, so that we never know when or where to expect them. The comet discovered by the Czech astronomer Lubos Kohoutek in 1973 was typical. It caused a great deal of interest, and was widely studied, but by now it has withdrawn into deep space, and it will not come back for 75,000 years or so.

In the year 1682 – that is to say, just within our 'revolution' period – Edmond Halley, friend of Newton, observed a bright comet, and calculated that it should have a period of about 76 years. He predicted that it would return in 1758, and it duly did so. Calculations proved that it had been seen regularly in the past, and was, for instance, identical with the comet which blazed forth in 1066 and caused alarm and despondency in the court of King Harold because it was regarded as unlucky. Halley's Comet was on view once more in 1835 and in 1910, and was back again in 1986. It remains the only naked-eye comet which we can predict, and I will have more to say about it later.

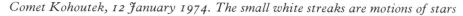

Comet Kohoutek, 12 January 1974. The small white streaks are motions of stars

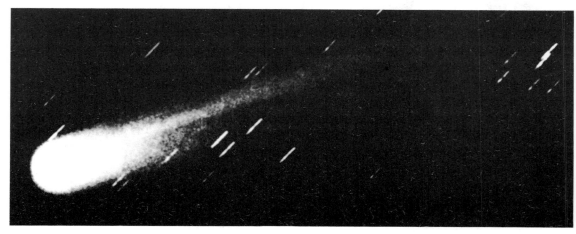

Because a comet is a long way from the Earth, it does not seem to move perceptibly against the background of stars. If you see an object travelling quickly across the sky, it certainly cannot be a comet. It may be an artificial satellite, of which hundreds have been launched since the opening of the Space Age on 4 October 1957 with Russia's never-to-be-forgotten Sputnik I, but if it is moving at high speed it is more likely to be a meteor.

A meteor is a tiny particle, usually smaller than a pin's head, travelling round the Sun in the same way as a planet. So long as it stays well clear of us, we cannot see it; but if it dashes into the top part of the Earth's atmosphere, it rubs against the air-particles and sets up friction, so that

Halley's Comet, 1066, from the Bayeux Tapestry

it destroys itself in the luminous streak which we term a shooting-star. When the Earth passes through a shoal of meteors (as happens, for instance, every August) we see a shower of shooting-stars. Now and then we encounter a larger body, which can survive the complete drop to the ground without being burned away, and is then called a meteorite. We now know that a meteorite is only a distant relation of a shooting-star, and is more closely associated with the asteroid belt; but to go into further details here would be pointless, and it is enough to repeat that there is no connection between a shooting-star and a real star.

To see a star at close range, you need only look at the Sun – because the Sun is nothing more nor less than an ordinary star, appearing much more splendid than the rest only because it is so much nearer to us. It shines by its own light, and it is huge. Its diameter is 865,000 miles, far

A shower of shooting stars, 9 October 1933

exceeding even Jupiter, and its volume is more than a million times that of the Earth. Yet we have learned that it is by no means distinguished, and on the stellar scale it is ranked as a dwarf. Its mass is, nevertheless, so great that the Sun is in complete control of its planet-family, and it sends us virtually all our light and heat, so that without it we could not survive for an instant.

The Sun is not burning in the accepted sense of the term; it is producing its energy by nuclear reactions, and this is true of all normal stars. It will not last forever, but it is unlikely to change much for several thousands of millions of years yet, so that we have no immediate cause to take to our rocket-ships and seek safety elsewhere!

In our own time, rockets have taken men to the Moon. Terrestrial isolationism ended at the moment when Neil Armstrong stepped out on to the bleak lunar rocks, in July 1969. Probes have by-passed the planets, even mighty Jupiter, and I have little doubt that if all goes well we will eventually be able to explore the whole of the Solar System. But when we consider interstellar travel, the situation is very different, because the stars are so remote. Even the nearest of them, beyond the Sun, lies at a distance of over 24 million million miles. Light, moving at 186,000 miles per second, takes $4\frac{1}{4}$ years to cover this gap, so that we say that the nearest star is $4\frac{1}{4}$ 'light-years' away.

The stars are suns in their own right, and many of them are much larger and more powerful than our Sun. We know of one star which shines as fiercely as a million Suns put together, though it is so remote that with the naked eye it cannot be seen at all. No matter what telescope we use, a star appears as a mere point of light, twinkling because it is seen through the unsteady layers of the Earth's atmosphere. The diameters of the stars cannot be measured directly, and when we study them we have to make use of more complicated and less direct methods of investigation. (If you turn a telescope toward a star and see a large, shimmering blob of light, you may be sure that there is some-thing wrong with the telescope.) It is only the members of the Solar System which can be examined in detail.

With the naked eye, a planet looks remarkably starlike; but from very ancient times it has been known that there is a fundamental difference. The planets wander slowly around the sky, whereas to all intents and purposes the stars do not—hence the old term of Fixed Stars. And the reason can be summed up in one word: Distance.

The farther away a moving object is, the slower it seems to go. A bird flying between the tree-tops appears to shift much more quickly than an aircraft seen against the clouds, even though its true speed is very much less. The stars are not genuinely fixed, and have extremely high velocities relative to the Earth, amounting to many miles per second, but they are so remote that to naked-eye observers they do not seem to shift at all relative to each other. The star-patterns seen by Newton, Galileo, Copernicus and even the Roman emperors and the builders of the Pyramids were to all intents and purposes the same as those of A.D. 1974. The groups or constellations described by the sky-watchers of thousands

Edwin Aldrin, the second man to walk on the moon, Apollo 11 mission, 1969

Sputnik 1, the first Soviet artificial earth satellite launched in 1957

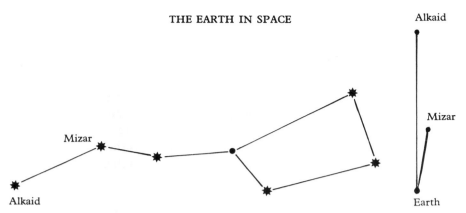

Figure 2: Alkaid and Mizar in the Great Bear.

of years ago gave every impression of being permanent, and it was quite logical to suppose that they never changed.

The 'wandering stars' or planets were quite different, and the most casual observer could see that they moved around from one constellation into another, though it is true that they always kept to certain well-defined regions. The Moon, too, was a quick mover, and it was tacitly assumed that both it and the planets were closer to us than the stars. Yet when you look at the sky there is no three-dimensional effect, and originally there was no means of knowing whether or not the stars were all at the same distance from us.

Modern research has shown that there is a very wide range indeed. Consider, for instance, two of the stars in the famous pattern which is known officially as the Great Bear and unofficially as the Plough or the Big Dipper. They have been named Alkaid and Mizar, and Alkaid looks slightly the brighter of the two. Though they lie side by side in the sky as seen from Earth, they are not true neighbours, as the diagram will show (Fig 2). Mizar is 88 light-years from us, Alkaid 210, so that Alkaid is considerably farther away from Mizar than we are. (Remember that one light-year is the distance which a ray of light will cover in a year; it amounts to rather less than six million million miles.) We are dealing with nothing more significant than a line-of-sight effect; and if we could look at the two stars from a vantage point somewhere between them, they would appear on opposite sides of the sky.

It follows that what we call a constellation is not a true group at all. The stars in any particular constellation need not necessarily have any association with each other. Moreover, the patterns are not permanent if we consider periods of time which are sufficiently long, because each star has its own tiny individual or 'proper' motion compared with its fellows. In the Great Bear, for example, two of the stars are moving through space in a direction opposite to the other five; if we could come

Photograph of full Moon

back to the Earth in, say, fifty thousand years the familiar shape would have become badly distorted.

To underline once more how big the star-system really is, a scale model may help. Consider a line one inch long, and take it to represent the Earth–Sun distance, which is–as we have noted–approximately 93 million miles. On this scale the nearest star will be four miles away; the distance of Alkaid will be over 200 miles, while the brilliant white star Rigel in Orion must be taken out to at least 900 miles. No rocket probe we can visualize as yet can have any hope of taking us upon journeys beyond the Sun's family. Interplanetary probes have become science fact, but interstellar probes still belong to the realm of science fiction.

The system of which our Sun is a typical member contains about a hundred thousand million stars all told. The Galaxy, as the system is called, is somewhat flattened in shape, and the Sun, with the Earth and all the other planets, lies well away from its centre. Look along the main plane of the system, and you will see many stars in roughly the same direction, as shown in the diagram. This produces the glorious band of light which we know as the Milky Way, and which has been the subject of many old legends, though nobody realized its starry nature until Galileo turned his telescope toward it in the winter of 1609–10.

With any modern telescope, or even a pair of binoculars, you too can see countless stars in the Milky Way, and it is easy to suppose that they are crowded together. This is not so. The stars are very widely separated, and they are in no danger of collision with each other. Neither is there any fear that a wanderer will suddenly appear from the depths of space and meet the Sun head-on.

There are stars of all kinds in the Galaxy. Some of them are huge and red, others white and fiercely hot; we know of stars which are variable, stars which explode, and pairs and groups of stars which make up genuine physical systems. In addition there are great clouds of dust and gas which we call nebulæ, and in which fresh stars are being born. A few of these nebulæ are visible with the naked eye; there is one, for instance, just south of the three bright stars which make up the so-called Belt of Orion. Neither must we overlook the clusters of stars, of which the best-known example is the Pleiades or Seven Sisters in the constellation of the Bull. The variety seems almost endless.

We cannot see through to the centre of the Galaxy, because there is too much dust in the way; it is rather like standing in Parliament Square and trying to read the dial of Big Ben on a foggy night. However, we do know that the centre lies in the direction of the magnificent star-clouds of Sagittarius, the Archer, and that it is approximately over 30,000 light-years away. The overall diameter of the Galaxy is of the order of 100,000 light-years.

Even this is only the beginning. Far away in space, so remote that their light takes millions of years to reach us, we can see other galaxies, each with its own quota of stars—and, no doubt, with its own quota of inhabited planets. Some of the outer galaxies are spiral in form, like Catherine-wheels, and modern research has proved that our Galaxy also is a spiral, with the Solar System lying near the edge of one of the

Star-clouds in Sagittarius

arms. We have also found that apart from the very nearest galaxies, all these 'star-cities' are moving away from us, so that the whole universe is expanding. Whether this spreading-out will go on indefinitely is something which we do not yet know, and neither have we any real idea of how the universe itself came into existence. All we can say with confidence is that it is thousands of millions of years old.

Ideas of this kind were quite beyond the mental grasp of the early star-gazers, and it was only with the work of Sir Isaac Newton that what we may call the modern phase of astronomy began. But this was at the very end of the great revolution; and having paved the way, let us now go back and begin at the beginning.

The Questioners

Early man believed the Earth to be flat. He also believed it to be stationary, and to lie in the very centre of the universe, with the sky revolving round it once a day. The stars were luminous points fixed on to a solid crystal sphere, and all the bodies in the heavens had been put there for the express benefit of humanity. Unusual phenomena, such as eclipses and comets, were signs of divine displeasure, and meant that the gods had to be placated in whatever manner seemed best.

Ideas of this kind sound ludicrous enough today, when every school-boy knows quite well that the Earth is a globe and that the sky is not solid. Yet the one thing we must not do, in looking back over the old theories, is to laugh. We have all the learning of the centuries at our disposal, and we have equipment of all kinds, ranging from powerful telescopes capable of picking up the light of objects thousands of millions of light-years away to the strange-looking devices which we call radio telescopes, used for studying long-wavelength radiations which are completely invisible to our eyes. Our knowledge has not been gained easily, but there has been plenty of time. The ancients had none of these advantages; and after all, what proof could they have that the Earth is either spherical or moving?

The landscape in general appears to be a flat surface, apart from local irregularities such as hills and valleys. Our ancestors had no means of travelling far from their homes, and they had no maps. They could see the sky turning, with the Sun rising toward the east and setting toward the west; why should it not be turning in reality? There was no reason to think otherwise. Also, they depended upon the Sun for their light and warmth, with a lesser contribution in light from the Moon, and this surely indicated that there must be some divine power regulating it or even residing in it.

Organized religion did, it is true, produce some ideas which are bizarre by any standards. The Egyptians believed the universe to be shaped like a rectangular box, with the longer sides running north–south. There was a flat ceiling, supported by four pillars at the cardinal points, and the pillars were linked by a chain of mountains, below which ran a ledge containing the celestial river Ur-nes. Along this river sailed the boats which carried the Sun and other gods, and there seemed to be no difficulty whatever in turning corners. In parts of the Nile Delta it was thought that the heavens were formed by the body of a goddess with the rather appropriate name of Nut, who was permanently suspended in a position which was as uncomfortable as it was inelegant. In India, the Vedic priests taught that the flat Earth was supported on twelve massive pillars, and that during night-time the Sun threaded its way between these pillars without bumping into any of them. And we also have a Hindu theory, according to which the Earth was carried on the back of four elephants standing on the shell of a tortoise which was itself supported by a serpent floating in a boundless ocean.

On the credit side, some remarkably accurate observations of star-positions were made, and the Egyptians produced a 365-day calendar. This was based largely upon the so-called heliacal rising of Sirius, the brightest star in the sky: that is to say, the date when Sirius could first be seen rising before the Sun at dawn. There were good reasons for making measurements of this kind. The whole Egyptian economy depended upon the annual flooding of the Nile, and it was necessary to know just when this flooding could be expected. Note also that the Great Pyramid, still dominating the landscape today even though it was built so long ago, is astronomically oriented.

In fact, Egyptian astronomy was a strange medley of careful, accurate observation and equally inaccurate, unscientific interpretation. There were, of course, many gods, but in or about the year 1379 B.C. the young Pharaoh Amenophis IV tried to make a radical change. He adopted the name of Akhenaten, and founded the cult of Sun-worship; there was now only one god, the Sun-disk or Aton, and the Pharaoh even set up a new capital city, from which he did his best to organize a religion based upon love and happiness. We still have the 'Hymn to the Sun' which he wrote, and which remains unique. Inevitably the experiment failed;

Opposite: *Philosophers and astronomers observing the heavens; from an early fifteenth-century manuscript*

Overleaf: *Northern constellations on the ceiling of the Tomb of Seti I in the Valley of the Kings, Luxor, 1303–1290 B.C.*

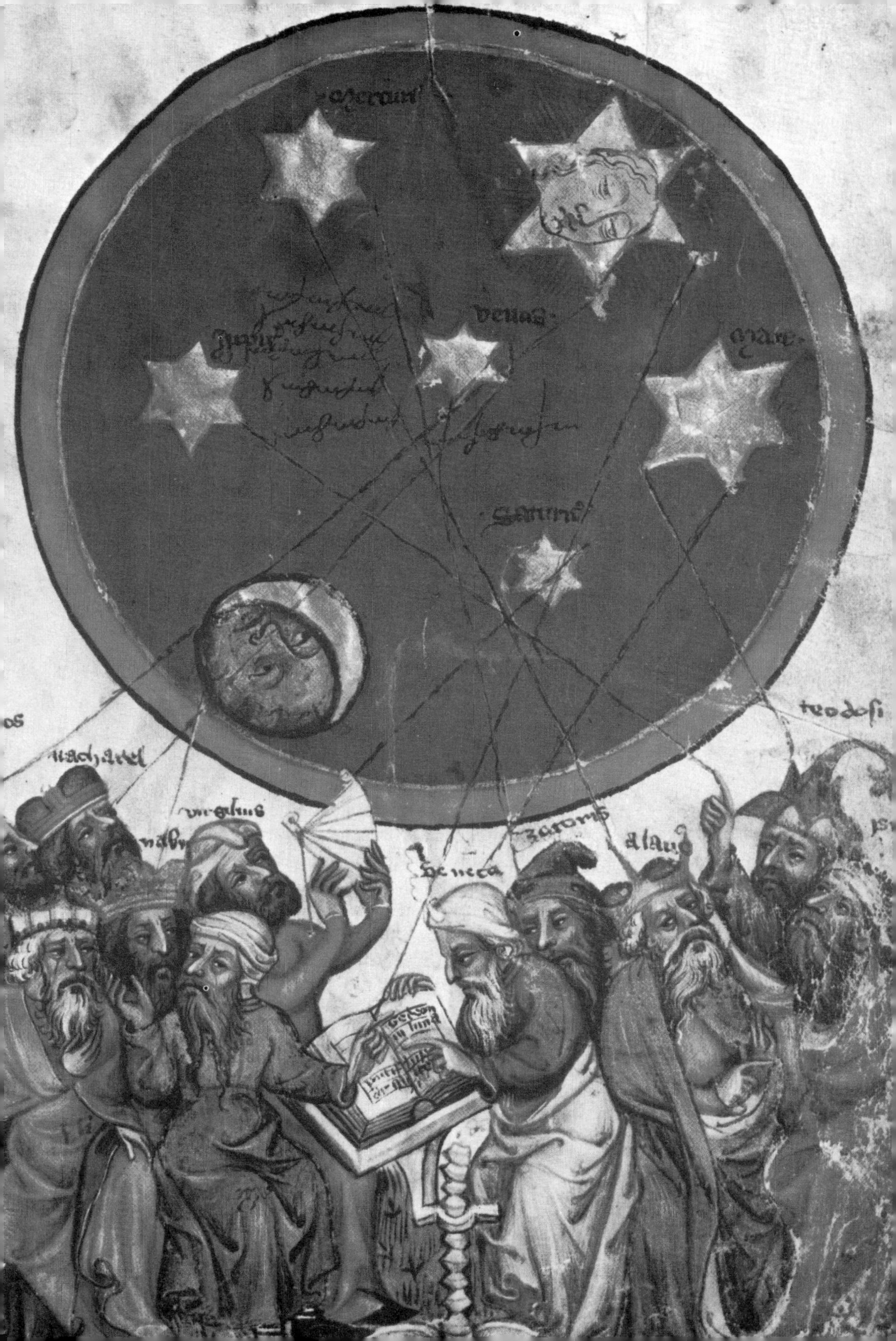
venus
virgilius
Seneca
teodosi

Prolemeus

Astronomi

Akhenaten was overthrown by the orthodox priests, and his city crumbled into ruin.

The other great civilization of very ancient times, China, was no more enlightened from the theoretical point of view, and I cannot resist quoting a famous story which may be true–though I admit to having grave doubts about it. Because the Moon is in orbit round the Earth, it must sometimes pass between the Earth and the Sun, cutting out the brilliant solar disk for a brief period and causing what we call an eclipse. The Chinese put this down to a hungry dragon which was trying to eat the Sun, and considered that the only remedy was to make as much noise as possible until the creature desisted (a procedure which, it may be added, always worked!). According to the legend, an eclipse took place in 2135 B.C., during the reign of the Emperor Chung K'ang of the Hsia Dynasty, but the two Court astronomers, who rejoiced in the rather peculiar names of Hi and Ho, failed to predict it–for which crime they were summarily executed.

Alas, the story is apocryphal, but it does indicate that the Chinese, too, were reasonably skilful observers. Eclipses tend to recur after an interval of just over eighteen years, or more precisely 6585 days 8 hours; this period is known as the Saros, and means that after one Saros the Earth, Moon and Sun return to almost the same relative positions. It is not exact, but it is a good approximation. The Greeks knew about it, and presumably the Chinese did too.

By the time of Thales, first of the great Greek philosophers (624-547 B.C. or thereabouts) the stars had already been divided up into constellations. The groups worked out by the Chinese were different from those of the Egyptians, which in turn were not the same as those of the Greeks. Had we followed, say, the Egyptian system our sky-maps would have looked very unlike those we actually use, though it is hardly necessary to add that the stars themselves would have been in the same places.

It is tempting to dwell further on these old ideas; there is much to be said, and I have not even mentioned people such as the Babylonians, who also made measurements of star positions. But it is time to pass on to what is often called the 'Greek miracle', because it is here that we find the first glimmerings of the great revolution which was to come so many centuries later.

Opposite: *Ptolemy holding a quadrant followed by Urania, the Muse of Astronomy. From Gregor Reisch's* Margarita Philosophica, *Basle, 1508*

It has often been supposed that the Greeks caused a quick, complete change in human outlook, and that from a standing start, so to speak, they developed a true scientific technology in an amazingly short period of time. Nothing could be further from the truth. Thales was born in 624 B.C., and Ptolemy of Alexandria, the last of the great philosophers of the Greek school, died in or about A.D. 180. This means that the whole story extended over more than eight hundred years. In time, Ptolemy was as remote from Thales as we are from the Crusades. Developments were not rapid; they were very slow indeed, and some of the early philosophers held ideas which were little more advanced than those of the Chinese or the Egyptians. Thales himself will always be remembered for his prediction of an eclipse of the Sun, which led to an abrupt halt in a battle between the Lydians and the Medes in the year 585 B.C., but he certainly believed that the Earth must be floating on water, rather like a log or a cork. There is no true science in this.

On the other hand, his younger contemporary Anaximander was bold enough to declare that the Earth is suspended freely in space, without support, because it is at an equal distance from all other celestial bodies and is therefore in a state of equilibrium. This was a definite advance, though his ideas in other respects were primitive. True, he taught that the Sun is equal in size to the Earth, but added that 'it is like a chariot-wheel, the rim of which is hollow and full of fire, and lets the fire shine out at a certain point in it through an opening like the nozzle of a pair of bellows'. Pythagoras, the great geometer, was the first to state definitely that the Earth is spherical and that the planets have independent movements in the sky–though the latter, at least, must have been known earlier. *En passant*, Pythagoras was also the founder of a sort of religious brotherhood, whose members were forbidden to poke a fire with an iron bar or to eat beans.

The great pyramids of Gaza, Egypt

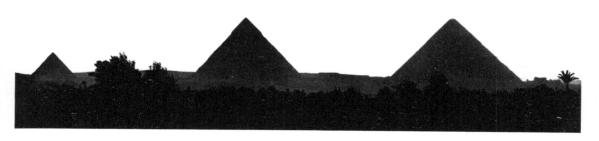

Xenophanes of Colophon, who was born about 570 B.C. and lived to be over ninety years of age, had his own theories. The Sun, Moon and stars were made of clouds set on fire, but when the Sun set at nightfall it was extinguished, to be replaced by a new one next day. 'There are many suns and moons according to the regions, divisions and zones of the Earth, and at certain times the disk falls upon some division of the Earth not inhabited by us, and thus when, as it were, stepping where there is void, exhibits eclipse.' It was at this time, too, that yet another philosopher, Heraclitus, taught that the Sun can be no more than twelve inches in diameter, which is something of an underestimate!

All this has been preliminary to the main story, but with the career of Anaxagoras, born about 500 B.C. at Clazomenæ (not far from Smyrna) we come to the first sign of persecution – and persecution was to play a vital rôle later. Anaxagoras soon moved to the cultural centre of the world, Athens, and became famous as a teacher; he once said that the object of being born was 'to investigate the Sun, the Moon and heaven'. He believed in a flat Earth, but he knew that the Moon shines by reflected sunlight, and this led him on to a correct explanation of eclipses.

*The Chinese thought of an eclipse as being a hungry dragon eating the Sun;
from a fifteenth century woodcut*

Unfortunately for himself, he also claimed that the Sun is a red-hot stone larger than the Peloponnesus, the peninsula upon which Athens stands, and he was strongly attacked on the grounds of impiety.

There may have been political reasons for the onslaught. Anaxagoras was a close friend of Pericles, the ruler of Athens, and when Pericles became unpopular some of the odium rubbed off on his associates. We do not know whether Anaxagoras was ever in real danger, but he certainly had to leave his home, and lived in exile until his death at the age of seventy-two.

This is no place to delve into the views of all the leading Greek philosophers, but I must mention Eudoxus of Cnidus (408–355 B.C.), who worked out a system of 'concentric spheres' to account for the movements of the planets. He knew that if the planets simply travelled around the Earth in perfectly circular paths at regular speeds, they would have uniform motion against the starry background; but this does not agree with observation. Their speeds vary, and at times they seem to

Total Solar eclipse showing a great prominence upper right

stop and 'go backwards' or retrograde for a while. This was a fact which came into great prominence later. His contemporary Heraclides supported the idea of a rotating Earth, but it was the great Aristotle who laid down many of the principles which were slavishly followed by almost all scholars for many centuries.

Aristotle, of course, was a man of many parts. During his lifetime, which extended from 384 to 322, B.C. he studied philosophy, politics, ethics, biology and much else, so that he was by no means an astronomer first and foremost; but he did write one book, known now by its Latin title of *De Cælo* (*On the Heavens*) which became accepted as the definitive and infallible work. To him the Earth was spherical, and he gave some genuinely scientific proofs. For instance, the sky does not seem the same when viewed from different parts of the world; the bright star Canopus rises over the horizon from Alexandria but not from Athens, and this can only be explained by assuming the world to be a globe. Next, he considered eclipses of the Moon, which he knew to be caused by the shadow of the Earth falling on to the lunar surface. The edge of the shadow is curved; therefore the Earth itself must be curved. Finally, Aristotle described his theory that everything must seek to return to its 'natural place'; thus a solid object will fall down, while a flame will rise. If all the material making up the Earth tends to fall toward the centre of mass, the result will be a spherical body.

And yet Aristotle argued strongly against any idea that the Earth could be in motion. His authority was tremendous, and to question it was regarded as unthinkable even as late as the Middle Ages. It may be largely for this reason that the bold ideas of later men such as Aristarchus of Samos made so little headway—which is a pity, because it was this obsession with a stationary Earth which held astronomy back for so long.

Aristarchus was undoubtedly one of the most brilliant of the Greek thinkers. By bad luck his most important book has not come down to us in its original form, and we have to depend upon second-hand accounts of it, though fortunately it was described in some detail by another great pioneer, Archimedes (who lost his life when he was working out a mathematical problem, and refused to pay due attention to a Roman soldier who was anxious to question him). Neither do we know much about Aristarchus himself, though he seems to have been born in or near 310 B.C. and to have lived to the ripe old age of eighty. To quote Archimedes in his *Psammites* or *Sand-reckoner*:

'But Aristarchus of Samos brought out a book consisting of certain hypotheses, in which the premises lead to the conclusion that the universe is many times greater than that now so called. His hypotheses

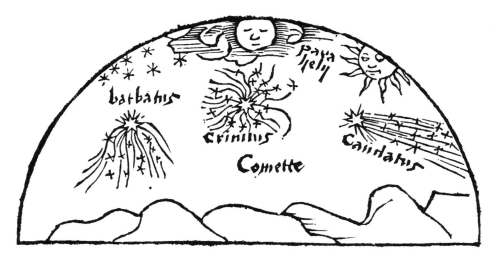

Woodcut of meteors and comets, from Aristotle's Meteorologia, *1512*

are that the fixed stars and the Sun remain motionless, that the Earth revolves about the Sun in the circumference of a circle, the Sun lying in the middle of the orbit, and that the sphere of the fixed stars, situated about the same centre as the Sun, is so great that the circle in which he supposes the Earth to revolve bears such a proportion to the distance of the fixed stars as the centre of the sphere bears to its surface.'

Re-write this in modern phrasing, and it is a clear statement that Aristarchus anticipated Copernicus by eighteen centuries. He also did his best to calculate the sizes and distances of the Sun and Moon by using methods which were perfectly sound in theory, though they broke down in practice because the measurements needed could not be made accurately enough. He knew the Sun to be larger than the Earth, while the Moon is smaller; and all in all, he put forward ideas which were far in advance of his time. According to the Roman historian Plutarch, there were vague proposals to accuse Aristarchus of impiety, but so far as we know nothing came of them.

Yet the idea of a moving Earth was too bold even for the Greeks. It was supported by a Chaldæan named Seleucus, who lived about a century after Aristarchus, but otherwise it was ignored, and after a while it was more or less forgotten except as an historical curiosity. Instead, cumbersome and artificial-sounding theories were produced, and were brought to their highest degree of perfection by Ptolemy at the very end of the Greek period. This is not to say that astronomy stood still: far from it. For instance, Eratosthenes of Cyrene, born about 270 B.C., measured the size of the Earth with amazing accuracy, and rather more than a century later Hipparchus of Nicæa not only drew up a really good star-catalogue but also improved upon Aristarchus' estimates for the sizes and distances of the Sun and Moon. He was also the first mathematician to make systematic use of trigonometry, and he discovered the effect which we call precession, due to the fact that the Earth's axis does not point in a constant direction; it shifts slowly but steadily, so that the position of the pole of the sky shifts as well.

I am well aware that this brief account of Greek astronomy is painfully sketchy and incomplete. Many important people, and many important theories, have been left out. But I am anxious to speed on to the sixteenth century; I have not set out to discuss earlier times, and in the B.C. period only Aristotle and Aristarchus are of true relevance to my main theme. So let me turn now to Claudius Ptolemæus, better known as Ptolemy – the 'Prince of Astronomers', who has left us an invaluable summary of the final state of Greek scientific thought.

We know absolutely nothing about Ptolemy's life or personality. His great book, which has come down to us under its Arab title of the *Almagest*, seems to have been written about A.D. 140, and the date of Ptolemy's death is usually given as A.D. 180, but we are not even sure of his nationality. He spent virtually all his working period at Alexandria,

Bust of Archimedes: Museo Capitalino, Rome

35

where, incidentally, there was an unique collection of books (Eratosthenes had once been in charge of the Library). Ptolemy's star catalogue was based on that of Hipparchus, and it has sometimes been claimed that he was little more than a copyist, but the evidence shows this accusation to be most unfair. He made notable advances on his own account, and he was also a geographer and a mathematician of the highest order. It was Ptolemy who drew up the first world-map which was based upon scientific measurement rather than inspired guesswork, and the Mediterranean area is shown in recognizable form; so for that matter is Britain, even though Scotland is joined on to England in a sort of back-to-front position.

The official theory of the universe, as described in the *Almagest*, is always known as the Ptolemaic, though Ptolemy himself did not invent it; he merely refined it and gave a comprehensive account of it. He had no patience with the absurd idea that the Earth might go round the Sun, and neither could he believe that the world is spinning round, because if the Earth were rotating beneath its atmosphere there would be a constant howling gale. Therefore, he had to follow a system which could explain the various wanderings of the Sun, Moon and planets without allowing the Earth to shift at all. Moreover, he was convinced that all paths of celestial bodies must be circular, since the circle is the 'perfect' form, and nothing short of perfection can be allowed in the heavens. (Remember my earlier comment that the slight eccentricity of the Earth's orbit is of the utmost significance.) So Ptolemy accepted a plan involving what are known as epicycles and deferents, drawing upon the work not only of Hipparchus but also, notably, of the earlier philosopher Apollonius.

Why not suppose that each planet moves in a circle round the Earth at uniform rate? Here, Ptolemy was faced with the same difficulty which had confronted Eudoxus of Cnidus so long before. A planet cannot move across the sky in an irregular fashion if it is in a set, unchanging circular orbit around us, and Ptolemy was under no delusions. Of the naked-eye planets, Mars, Jupiter and Saturn describe 'loops' over periods of weeks; Mercury and Venus have their own way of behaving (Heraclides, around 350 B.C., had even suggested that they travel round the Sun, though he believed the Sun to orbit the Earth) but the wrong-way or retrograde movements had to be accounted for somehow. According to the Ptolemaic theory, a planet moves round the Earth in a small

Opposite: *Clerk, Astronomer (with astrolabe) and Mathematician, from an early thirteenth-century* Psalter of St. Louis and Blanche of Castille

circle or epicycle, the centre of which – the deferent – itself moves round the Earth in a perfect circle. Even this did not do, and more and more complications had to be introduced. This is what I meant by describing the whole plan as cumbersome (Fig 3).

And yet it was by no means illogical in view of the state of knowledge two thousand years ago. Nothing was known about the laws of gravity, and there was no serious thought that the planets could be worlds of the same basic nature as the Earth. There is one vital point to be borne in mind. Given that the Sun, Moon and planets could in fact behave in the way that Ptolemy expected, *the theory worked*. It explained the various wanderings, including the retrograde motions, and there was no obvious fault with it. It fitted in with accepted science as well as with accepted religion, so that everybody was satisfied with it (Fig 4).

With Ptolemy we come to the end of an era. Much had been achieved, and some golden opportunities had been missed, which was by no means surprising. The succeeding centuries were barren of ideas, and, tragically, the books in the great Alexandrian Library were lost. There is a legend that they were deliberately destroyed on the orders of an Arab caliph, Omar, who decreed that if the books contradicted the Koran they were heretical – while if they agreed with the Koran, they were superfluous. This seems highly suspect; more probably the books were simply dispersed, but in any case they have not come down to us, and the loss can never be made good. The only piece of luck was that Ptolemy's great work reached Baghdad by the eighth century A.D., and was translated into Arabic. Without it, our knowledge of the astronomy of the Classical period would be horribly slender.

It was, indeed, the Arabs who were responsible for the re-birth of astronomy. Their original motives were not purely scientific. At that time there was no clear-cut distinction between astronomy and astrology, the pseudo-science which attempts to link the positions of the Sun, Moon and planets with human character and destiny. All the old astronomers, including Ptolemy, were astrologers as well, and the cult remained powerful until surprisingly recent times. (In some countries, such as India, it still remains a force to be reckoned with.) The Arabs

Opposite top: *Angels turning the Planetary Spheres, which, according to the Ptolemaic system, carried the heavenly bodies round the motionless Earth. From a fourteenth-century Provençal manuscript*

Opposite bottom: *Nicole Oresme, a mathematician, with his armillary sphere. From his commentary on Aristotle's* De Coeli et Mundo *written in 1377*

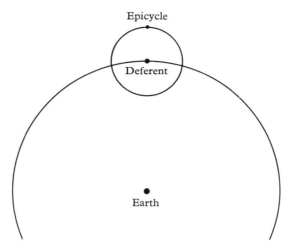

Figure 3: Epicycle and deferent.

of the ninth to fifteenth centuries A.D. were astrologically-minded, and of course this meant that they had to have good star catalogues as well as tables which would predict the movements of the planets. In astrological lore, everything depends upon the exact positions of the bodies of the Solar System against the background of the fixed stars.

At first the Arabs used the star catalogue in the *Almagest*, but as time went by they drew up catalogues of their own. They even built observa-

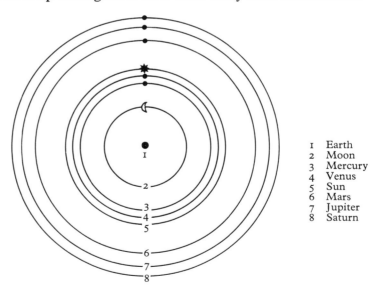

1	Earth
2	Moon
3	Mercury
4	Venus
5	Sun
6	Mars
7	Jupiter
8	Saturn

Figure 4: The Ptolemaic Universe.

tories. Telescopes still lay in the future, but the observatories included various measuring devices which were put to good use. For a while Baghdad became the astronomical centre of the world, and some excellent results were produced; incidentally, most of the individual star-names which we now use are of Arabic origin. In the year 1270 King Alphonso X of Castile called a number of Arab and Jewish astronomers to the city of Toledo, and was responsible for the publication of the famous *Alphonsine Tables*, which contained data for predicting the positions of the planets as well as telling when eclipses of the Sun and Moon were due. Then, in 1420, Ulugh Beigh, a Mongol prince who was a grandson of the Oriental conqueror Tamerlane, set up an elaborate observatory at his capital of Samarkand, and undertook a long programme of accurate scientific work. His star catalogue, in particular, was much better than Ptolemy's. Unfortunately for himself, he sent his eldest son into exile on astrological advice, with the predictable result that the son rebelled and had Ulugh Beigh murdered. That, to all intents and purposes, was the end of the Arab school of astronomy; and I do not propose to discuss it further, because no Arab astronomer even dreamed of questioning the Ptolemaic theory.

Meantime, astronomy was coming back to Europe. The spread of knowledge was made much easier by the invention of printing, and by the middle of the fifteenth century we find regular astronomical information being published from Nürnberg by one Johann Müller, better remembered by his Latinized name of Regiomontanus. Once again European thinkers turned their attention back toward the skies, and it was only a matter of time before bold new theories would be put forward.

Actually, the first mild revival of the old heliocentric idea was due to Nicolaus Krebs, the son of a German wine-grower, who was born in 1401 and died in 1464. He is often known as Nicholas of Cusa, since his birthplace was Cues on the Moselle. On the whole he is a rather shadowy figure, but he deserves to be remembered for a book, *De Docta Ignorantia*, in which he dared to suggest that the Earth might not be lying at rest in the centre of the universe. He took matters no further, but the first stirrings of revolt were there.

The stage was set for Copernicus.

Nicolaus Copernicus; an early sixteenth century woodcut by Tobias Stimmer

Copernicus, the Doubter

Toruń, on the banks of the river Vistula in Poland, is a pleasant city. It is neither very large nor very small; the Old Town is quaint and cobble-stoned, contrasting sharply with the massive University building and the modern hotels. Over the whole place there hovers the spirit of Copernicus, for it was here, on 19 February 1473 according to the old calendar, that the great astronomer was born. Toruń is justly proud of him, and so are the Polish people. There have been long discussions as to whether the Germans have any claim to him; but the evidence shows that he belongs properly to Poland, and the original spelling of his name was Mikołaj Kopernik. 'Nicolaus' (or Nicholas) 'Copernicus' is how we always refer to him, and in those days, of course, the Latinization of names was a common practice.

Let me add, too, that I propose to use the Polish versions of other names which we will encounter during the next few chapters: Frombork rather than Frauenburg, Toruń rather than Thorn, Łukasz rather than Lucas, and so on. Note also that the city we now call Gdańsk is the ancient Danzig, and that during the fifteenth century the capital of the country was Cracow; Warsaw did not yet exist.

In one way Copernicus was fortunate. Poland, as we all know, has had a troubled history; there have been long spells when it has been split up and dominated by foreign oppressors, but at the time of the great scientific revolution it was going through one of its most powerful and prosperous periods. The tragedy of partition did not come again until later. Not that there was anything in the way of blissful peace; there was constant trouble from the notorious Order of Teutonic Knights, and at one stage in his career Copernicus became heavily involved. But on the whole, Poles can look back at the whole period with a feeling of satisfaction. Learning and culture flourished, and Polish scholars were known all over the civilized world.

43

Also, Copernicus came of a family which was reasonably wealthy and certainly influential. Apparently his grandfather, Jan Kopernik, was a Cracow merchant, and it was here that Copernicus' father—also named Mikołaj—was born, though he moved to Toruń in or about 1458 and spent the rest of his life there. He too was a merchant, and Toruń was

Toruń, where Copernicus was born, from a plan published in 1540

an excellent centre for his operations, since at that time it had wide connections with many other towns and ports of Europe; to make things even easier he enjoyed the confidence of the Polish king, and served him loyally and well. Around 1460 he married Barbara, daughter of a wealthy alderman named Łukasz Watzenrode. The Watzenrodes were among the most celebrated citizens of Toruń, and they also had played a notable part in the defence of their country.

This brings me on to the story of the Teutonic Knights, or, as they preferred to be called, the Order of the Hospital of the Holy Virgin Mary of the German House in Jerusalem. The Order was founded in the twelfth century, in Palestine, out of the brotherhood which served in the German hospital for crusading knights, and within a few years it had turned itself into a sort of religious-cum-military organization, with the aim of both tending the sick and fighting against those who refused to accept Christianity. This sounds very high-minded and noble, but unfortunately the Order did not live up to its lofty ideals. In fact, not to mince matters, it became an unmitigated nuisance. Its members were recruited from German knights of noble birth, and habitually wore white cloaks marked with a black cross. When the Crusades showed signs of petering out, the Teutonic Knights remained very much of a force to be reckoned with, and those who called them into service usually had cause for regret. For instance, in the early thirteenth century a King of Hungary invited them to help in the defence of his realm, but drove them out when he realized that they were planning to set up their own independent state.

Meantime, the old Polish kingdom had disintegrated into a number of small duchies, each of which was too weak to protect itself against invasion. In 1226 Conrad I, ruler of one of these duchies—Mazovia—made the same mistake as the Hungarians had done, and called in the Knights to act as a shield against raids by the neighbouring Pruthenians, a pagan tribe of the Baltic group. The inevitable result was that the Pruthenians were slaughtered instead of being converted to Christianity, and the Knights succeeded in establishing what was virtually a State subject to nobody except themselves. Next they overran the whole province centred round Gdańsk, and they reached out to Toruń, where they built a massive fortified castle.

The Knights were not mild rulers. They were harsh and greedy, and no citizen was safe from their tax-collectors. It is not surprising that they were unpopular, and they had to be constantly on their guard. Toruń was selected as a fortress outpost against the Polish kingdom, and the castle loomed over the city as a symbol of darkness and oppression.

To be effective, resistance has to be organized, and there appeared a secret group with the rather curious name of the Salamander Society, which turned into the more stable Prussian Alliance. The result was a war which lasted for thirteen years from 1454, and it is now that we come to the first connection with Copernicus, because Łukasz Watzenrode – grandfather of Barbara, the astronomer's future mother – took an active part. Not only did he provide a large sum of money, but he also ventured on to the battlefield, and during one skirmish he was wounded, though apparently not seriously.

In February 1454 the senior citizens of Toruń informed the Grand Master of the Order that they were breaking all their ties with the Teutonic Knights. The guns roared out; the hated castle was attacked, and then taken by storm. Little remained of it by the time that the fighting ceased, and even before the smoke had ceased to swirl upward from the ruins a message had been sent to the Polish King, Casimir, asking that Toruń should be taken under his protection. Casimir was quick to agree, and in May he arrived in the city. 'Having ascended the royal throne, set up in the market-place and magnificently decorated, and wearing the mantle and all the insignia of royal authority', he accepted oaths of allegiance. For the moment, at least, the shadow had been lifted, and when the war came to an end the Knights lost many of their territories, including Gdańsk. Another region returned to Poland was Warmia, where Copernicus was to spend so much of his life.

Obviously the Teutonic Knights were not happy with the outcome, and they were determined to recover everything that they had lost. They were supported not only by the Germans but also by the Pope, and the peace was uneasy. Yet at least things had calmed down, and it was during this period that Nicolaus Copernicus was born.

His parents lived at 17 Anna Street, in Toruń (now, fittingly, called Copernicus Street) and Nicolaus was one of four children. The two girls do not really come into our story; Barbara became a nun, while Katarzyna married a merchant, living – so far as we know – a placid and contented life. But there was also Andrzej, three years older than Nicolaus, who shared many of his earlier experiences. Whether the boys were alike in character and ability we do not really know, but their careers were very different. Nicolaus achieved immortal fame, and lived to be over seventy; Andrzej is a tragic figure, who died deserted and alone. But this lay in the future, and at first there was no hint of what was to come.

Mikołaj – the father of the boys – had also taken part in the struggle against the Teutonic Knights, and his wealth and influence increased steadily. In 1480 he moved to a new house, number 36 in the Market Square (which, alas, was pulled down in our own century; the site is now occupied by a large department store) and all seemed well, but three years later he died, and we have every reason to believe that young Nicolaus and Andrzej walked sadly in his funeral cortège. The one comfort was that there was no immediate shortage of money, because Łukasz, Barbara Kopernik's brother and therefore Nicolaus' uncle, took a hand. He had been born in 1447 – do not confuse him with his father, also named Łukasz, who had been wounded in battle against the Teutonic Knights – and he was destined to play an all-important rôle in the lives of his two nephews. In his youth he studied at Bologna, and became a Doctor of Laws, but he followed the ecclesiastical path, and later became Bishop of Warmia, which meant in effect that he was more or less an independent ruler of a minor state. At the time of Mikołaj's death, Łukasz was already canon of Frombork, a small

The Polish Army besieges Malbork Castle, the seat of the Grand Master of the Teutonic Knights during the wars of 1454–67

47

cathedral town in one of the more remote parts of Poland, and he was well on the way to the highest honours.

Uncle Łukasz seems to have been a curious character. He was intensely ambitious, and could be ruthless when the occasion demanded. He was not renowned for his sense of humour, and it has even been said that he was never known to smile. Yet so far as his nephews and nieces were concerned his behaviour was always impeccable; he took them into his care, financed their education, and gave them all possible help.

When he took charge, Andrzej was nearly thirteen and Nicolaus ten. They had grown up in the fascinating city of Toruń, sometimes nick-named the 'Queen of the Vistula', with its narrow, crowded streets and its multi-lingual throng; they had explored the picturesque riverside, quays and the ornate buildings, and they had talked with merchants and traders who had come from all over Europe. Undoubtedly it was here that they attended their first school–probably St. John's, where for a brief period in his younger days Łukasz had himself been a teacher. Later they may well have gone to a cathedral school for older boys. But this was only a preliminary; higher education was essential, and the choice was fairly obvious. Łukasz was a graduate of the University of Cracow, and it was here that he decided to send his nephews. In 1491, when Nicolaus was eighteen, the brothers set out. The future was full of hope.

Cracow is no longer the chief city of Poland, but for charm and beauty it stands alone. There is no comparison with the modern capital of Warsaw, which is, after all, a completely new town inasmuch as little of it remained after the destruction wrought by the Nazis less than thirty years ago. Cracow takes us back into the graceful past. Overlooking it is Wawel Castle, once the residence of Polish kings such as Casimir the Jagiellonian, who had been reigning for forty-four years when Nicolaus and Andrzej began their studies. Wawel is utterly different from the castle of the Teutonic Knights which had cast a shadow over Toruń for so long. It is a symbol of greatness rather than gloom, and it is still impressive today, though admittedly it has been drastically modified over the centuries.

The Cracow of 1491 was famous not only in Poland, but throughout the civilized world. It was an administrative and trading centre, and it was also the seat of one of the most famous universities in Europe. The Jagiellonian University, as it is now called, was founded as early as 1364 by the then monarch (another Casimir), and was renowned for its teaching, so that students flocked to it from many other countries. Let me quote from the *Chronicles of the World*, by Schedel of Nürnberg:

Woodcut of Cracow, from the Chronicles of the World *by Schedel of Nürnberg*
1493

'There is in Cracow a famous university, which boasts many most eminent and highly-educated men, in which all sorts of proficiencies are practised, such as the study of speaking, poetics, philosophy and physics. But the science of astronomy stands highest there, and in all Germany there is no school that would be more renowned, as I know from the accounts of many persons.'

From the beginning of the fifteenth century the University had had one chair of mathematics and astronomy, and later another was added. When the Kopernik brothers arrived, the staff included several eminent astronomers, notably Wojciech of Brudzew (often called Brudzewski) and Jan of Glogów, both of whom were known far beyond the boundaries of their own country. Now, it is important to note that Nicolaus and Andrzej had not been sent to Cracow to learn about the stars. They were enrolled in the Faculty of Arts, and their studies were to be comprehensive. Neither is there any evidence that Nicolaus had been attracted to astronomy during his schooldays, though it is true that our knowledge of his boyhood is very slight. But on anyone with an inquiring mind, the influence of a teacher so skilled and persuasive as Wojciech was bound to have a tremendous effect, and the brothers were by no means immune to it.

University life in those days was varied and colourful. Some of the customs were strange, and in particular there was the initiation ceremony which all newcomers had to endure. The hapless candidate was treated in the manner of a block of wood, and so he had to be smoothed out – that is to say, planed with a wooden axe and then shaved with a huge razor, after which he was made to bath in dirty water. Then he was required to prove that he could read and write, though he was given a pen that would not function and an inkpot which defied all attempts to open it! If the judges felt that the tests had been passed, the victim (by now somewhat breathless, one may imagine) was given the privilege of treating the entire company to a feast. Only then was he allowed to rank as a fully-fledged member of the student community.

Nicolaus Copernicus was serious by nature, and we have no proof that he was subjected to the ritual, but he may well have been. At any rate, he was duly enrolled, and in the documents still in existence we find the entry: 'Nicolaus Nicolai de Torunia solvie totum' – in other words he had paid his fees, and was entitled to begin his studies.

Copernicus had been given the opportunity to learn, and he accepted it eagerly. No doubt he attended many lectures, during which the professor stood on a platform while the students sat either on wooden benches or – if benches were in short supply – on the straw-covered floor.

Later, the lectures would be reviewed under the supervision of senior tutors. Incidentally, the professors themselves were not exempt from University discipline, and if they failed to complete a lecture in the allotted time they were liable to fines or, in extreme cases, total loss of salary. (One can easily imagine the outcry if such a threat were made today.)

By this time Uncle Łukasz had become Bishop of Warmia, and was a very important man indeed, so that at Cracow his nephews were highly respected. Nicolaus made many friends with whom he maintained contact for the rest of his life; there was, for instance, Bernard Wapowski, later to become famous as a geographer and historian as well as in other branches of science.

Yet the most important result of those years at Cracow was the call of astronomy. Of course, all teaching was on the strictly conventional pattern; the Earth in the centre of the universe, with the Sun, Moon and planets moving round it in a complicated system of epicycles and deferents, with the star-sphere lying beyond. There are vague suggestions that Wojciech, in particular, was somewhat dubious, but he was not confident enough or bold enough to say so in public even if this is true. What, then, of Nicolaus Copernicus? He was a deep thinker, and even at Cracow, when he was still so young, he was not inclined to take

Łukasz Watzenrode, Bishop of Warmia and uncle of Copernicus

anything on trust. Scientifically he was a doubter, and there is every reason to assume that it was not long before he became uneasy about the whole Ptolemaic system. It was not straightforward, and it did not appeal to him.

He may have begun to formulate his own revolutionary theories at the time when he was studying at the Collegium Maius, the building at the University in which he spent so much of his time. If you stand there and look down on to the courtyard, as I did in 1973 when the International Astronomical Union held its celebrations in honour of the quincentenary of Copernicus' birth, it is easy to picture him: a serious, rather reserved young man, absorbing all the knowledge which his professors could give him, and pondering upon the many puzzles which seemed so hard to solve. There is a tranquil, timeless atmosphere about the Collegium Maius. Visit it, and you will soon understand what I mean.

All good things must come to an end, and in the autumn of 1495 Nicolaus Copernicus had to leave Cracow. There is some doubt as to whether he had taken any degree—on the whole, it seems that he had not—but he had certainly benefited to the full, and the Jagiellonian University had served him well. Today there are many relics of his stay there, and he remains the most famous of all Cracow students.

Meantime, important things had been happening in Poland. King Casimir had died in 1492, and Łukasz Watzenrode—now, you will recall, Bishop of Warmia—had become a trusted adviser of the new king, John Albert, who actually visited the University during Copernicus' spell as a student. Three years later Jan Zanau, canon of Frombork, also died, and Uncle Łukasz was anxious for his nephew to assume office. True, Nicolaus was only twenty-two, but this was no real obstacle (it is on record that a boy of fourteen was once appointed a full cardinal), and the omens seemed good. Yet for once in his life the Bishop met with a rebuff. Powerful though he had become, he had his enemies, and he had to admit temporary defeat. Being the sort of man he was, he simply bided his time, and for the moment there was no reason why his nephews should not continue with their education. The next step was the University of Bologna, in Italy, where Łukasz had taken his Doctorate of Laws so many years before. And so Nicolaus was again sent on his way.

The Wandering Scholar

The University of Bologna has the distinction of being the oldest in all Italy. It had been founded in 1119, and was particularly renowned for its Faculty of Law. This, obviously, is why Copernicus was sent there; astronomy was an interesting hobby, of course, but so far as Łukasz Watzenrode was concerned it was nothing more, and his main idea was for his nephews to make rapid progress in the Church. There were, incidentally, connections between Bologna and Poland. Between 1448 and 1471 there had been no less than five eminent Polish professors occupying chairs of astronomy and mathematics at the University.

The Italy of the late fifteenth century was far from a peaceful place, and religious toleration was conspicuous only by its absence. Yet there were plenty of brilliant scholars, and the University of Bologna was crowded with men of lively intellect. We may imagine that Nicolaus Copernicus was very much at home there, particularly as he soon struck up a close friendship with a leading professor of astronomy, Domenico Maria Novara, and may even have taken rooms in the professor's house. The problems of the Sun, Moon and planets were never far from his thoughts.

I describe Novara as a professor of astronomy, but this was only part of his duty. Officially he held the chair of astrology, and the two offices were combined. It is strange to remember that astrology was still classed as a true science less than five centuries ago, and it makes us realize, too, that Copernicus himself belongs to a different era. It is Galileo who may be regarded as the 'link-man' between ancient and modern.

But this is running ahead of our story. Copernicus had become a student of canon law in Bologna, and no doubt he worked hard at his studies. In fact, a student had to get up early; the first lecture might be timed for seven o'clock in the morning, and could last for two or three hours. There is a note in the old University rule-book that students were forbidden to start 'drumming on their benches' to indicate impatience. On the other hand a lecture would not, in general, overrun its allotted time. If it did, then the professor in charge could be reprimanded or fined, as at Cracow.

At Bologna, Domenico Novara was the man who had most influence on the still-youthful Nicolaus, and it was on 9 March 1497 that the two made an observation of an event which took place in the sky. As the Moon moves along, there are times when it passes in front of a star, hiding or occulting it. One of the really bright stars which can be occulted is Aldebaran, the orange 'Eye of the Bull', and this was what the professor and the student saw. It is, in fact, the first recorded astronomical observation by Copernicus, though certainly not the last.

Did Novara himself have any doubts about the truth of the Ptolemaic theory, and did he have the slightest faith in the wild idea that the Earth might be in motion? If he had, then he was too prudent to say so; heresy was even less safe in Italy than in Poland. Also, everything in astrology is based upon a central Earth, and it was no part of the duty of an astrological lecturer to start asking awkward questions. But from what happened later (and not so very many years later) there is every reason to think that Copernicus' own doubts were becoming more and more marked, and the observation of the occultation of Aldebaran may have strengthened them. Novara himself wrote several books, mainly about astrology, though unfortunately they have been lost.

Bologna, like Cracow, still honours Copernicus as one of its most distinguished University students, but we cannot pretend that we know a great deal about his stay there. He did not take his degree in canon law, and apparently he spent a large part of his time in studying not only astronomy, but also Greek. (Later, he published a Latin translation of some verse by the seventh-century Greek poet Theophylactus, which is the only reason why Theophylactus is ever remembered today.) Meantime, Andrzej had joined Nicolaus at Bologna, and the two brothers were not aloof from the usual kind of student life. It is on record that they once ran out of money, so that Andrzej considered going to Rome and taking the first job there that he could find. Fortunately Bishop Łukasz' secretary happened to be in Bologna at the time, and funds were supplied.

A second attempt to install Nicolaus as Canon of Warmia was suc-

Opposite: *Nicolaus Copernicus : anonymous portrait*

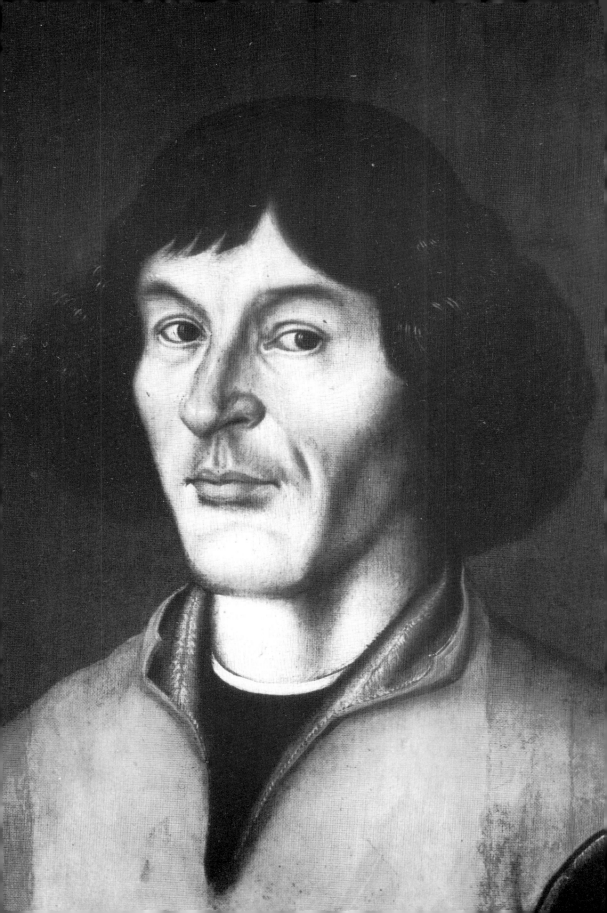

t huitieſme liure le quel parle
outes les planetes ot
double mouuement
dont lun leur eſt na
turel et propze qui
dotent en ozidént encontre le mo
ment du firmanīt lautre eſt vn
mouuement eſtrange qui eſt de onēt
ozidént par le firmaniēt qui les
enſt chaſain tour des le leuer uil

du double mouuemēt des
nettes et chaſaine en general
ment naturel ou quel elles leſt
deler coutre le firmamēt. Nuā
des planetes parfont leurs rou
pluſtoſt et les autres pluſtart
eſt pour ce que la quantite de
cercles neſt pas egale lune a la
Car ſaturne deuieure en chūn
par rro mois et aconpliſt ſō
en xxx ans Iupiter deuieure

cessful. This was probably in 1497, and two years later Andrzej also obtained a Canonry. This did not mean that they had to return at once; they were given leave of absence, and in 1500 we find them in Rome, the very centre of the Christian faith.

The year 1500 was celebrated in a grand manner: fifteen centuries of Christianity! And yet one has to admit that the Vatican, at least, was going through one of its less admirable periods. The Pope, Alexander VI, was one of the notorious Borgia family, and neither he nor his children, Cesare and Lucretia, could be regarded as models of virtue. Not to put too fine a point on it, they were scoundrels of the first magnitude, and there is little doubt that the Pope personally ordered the poisoning of several candidates for the Papal throne, while the best way to become a cardinal was by open bribery. Yet almost a quarter of a million pilgrims from the whole of the Christian world came to the Eternal City in that year of 1500, and Nicolaus and Andrzej were among them. We have definite proof that Nicolaus was there on 6 November 1500, because he observed and recorded an eclipse of the Moon.

Moreover, he had become sufficiently well-known to be invited to give lectures on astronomy. According to Rhæticus, who was to play so major a part in the publication of Copernicus' great book so long afterwards, the young scholar 'lectured in Rome, with large audiences and in a circle of eminent men' on astronomy. It is a great pity that we do not know just what these lectures contained. They can hardly have been in open contradiction to the Ptolemaic theory; this would have produced unfavourable comment at best, and persecution at worst. But with every passing year, Copernicus' qualms were increasing.

Obviously, the brothers had to go home to be officially installed in the Cathedral Chapter at Frombork, and so in May 1501 they left Rome and set out for Poland. On the way to Frombork they stopped at Toruń, to find that the quiet city had been turned into what was to all intents and purposes an armed camp. King John Albert, Casimir's successor, was not an energetic king, but he had been forced to realize that sooner or later the Teutonic Knights would have to be dealt with; their belligerancy had increased, and they were not inclined to give up their claims to the territory they had lost after the Thirteen Years' War. Cavalry regiments were passing through Toruń, and the King himself was there, so that Nicolaus and Andrzej certainly saw him. But John Albert died

Opposite: *Earth, the planets, the Sun and Moon surrounded by the twelve astrological signs. From a fifteenth-century French manuscript* Le Propriétaire des Choses

suddenly in the following June, and the confrontation with the Teutonic Knights was delayed for more than another decade.

Back in Frombork, the brothers appeared before the Chapter. The ceremonies were duly completed, and the two immediately requested further leave of absence, so as to continue studying in Italy. There seemed no reason to object, particularly as Nicolaus expressly stated that as well as completing his degree in canon law he wanted to study medicine – a subject in which he was already remarkably well-informed. This suited the Chapter very well, because doctors were scarce and medical science was remarkably primitive. Nicolaus was instructed to learn enough to make him 'a physician salutary to our Reverend Lord' (i.e. the Bishop) 'and to the gentlemen of the Chapter'. So for the last time the brothers left Poland, Nicolaus heading for Padua and Andrzej for Rome.

Padua, too, was a famous University, and since it dated back to the year 1222 it was second only to Bologna in seniority. It had a well-known School of Law, but was most celebrated for its Faculty of Medicine. Nicolaus completed his law studies there, but for some reason or other he elected to graduate at another university, that of Ferrara. It has been claimed that he chose Ferrara because it was there that the examination fee was lowest; at any rate, he became a Doctor of Canon Law on 31 May 1503.

However, much of his time was spent in medical studies, and although he never took an official qualification he became an excellent doctor by the standards of the time. Some of the training was carried out under conditions which would hardly be acceptable today. Back at Cracow, for instance, there had been a tradition that when a public execution was due to take place in the Market Square, the whole of the Medical Faculty, including the Dean, would don official robes and attend the ceremony, making careful notes while the luckless victim was drawn and quartered. (When there were no convenient executions, the Faculty would go instead to the local slaughterhouse – presumably without bothering to put on their robes – and use pigs as substitutes.) At Padua there was a full course lasting for three years, and Copernicus threw himself into his studies with his customary energy. We still have some of his books, and in the margins are some notes written in his own hand. I give a selection of comments and prescriptions, though I hasten to add that I have not personally tried any of them, and would be rather reluctant to do so:

Opposite: *The anatomy theatre at the University of Padua, most celebrated for its Faculty of Medicine*

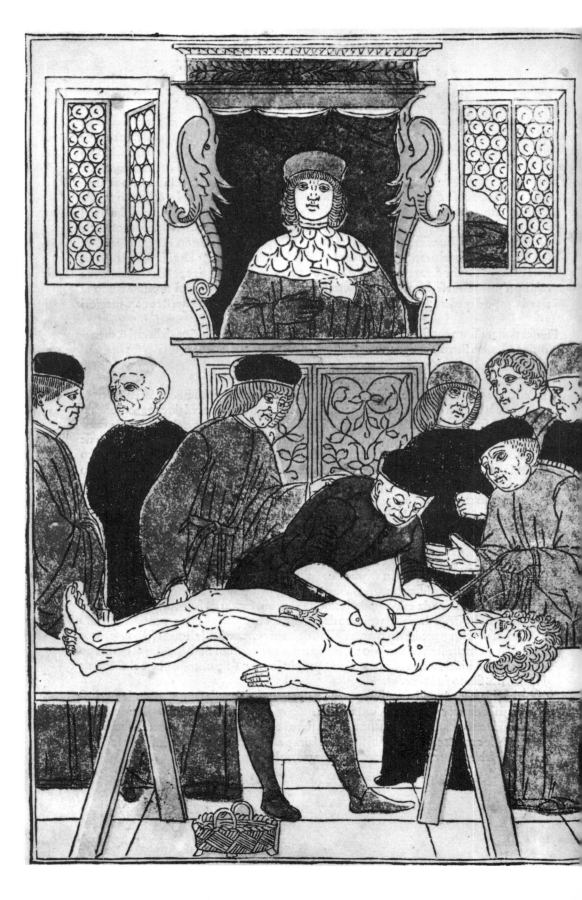

'Take resin from a fruit tree, boil it in beer three times, and drink with meals: helps podagra.' (Podagra, incidentally, is gout in the feet.)

'A remedy against the paralysis of the body: make an infusion of sage leaves, add rue, castoretum, boil in wine and drink.'

'A note on corns—look in the Pandects about the weeping willow.'

'For a remedy to protect one against the bite of a rabid dog, see the Pandects under the word *sapphire*.'

'Water extracted from beech leaves helps everything.'

And if you want a kind of universal remedy:

'Take two ounces of Armenian clay, a half ounce of cinnamon, two drachms of tormentil root, dittany, red sandalwood, a drachm of ivory and iron shavings, two scruples of ash and rust, one drachm each of lemon peel and pearls; add one scruple each of emerald, red hyacinth and sapphire; one drachm of bone from a deer's heart; sea locusts, horn of a unicorn, red coral, gold and silver foil—all one scruple each; then add half a pound of sugar, or the quantity which one usually buys for one Hungarian ducat's worth.' Significantly, he added: 'God willing, it will help.'

Copernicus also makes the cheerful comment that 'those who inherit diseases are rarely cured of them, and will be wise to endure their suffering in patience.' Yet despite all this, there can be little doubt that Copernicus was more skilled than most medical men of his time, and later, when he was back in Poland, his fame as a doctor became greater than his fame as an astronomer so far as his contemporaries were concerned. In 1519, for instance, there was a serious epidemic in Warmia, and it is said that Copernicus' treatment saved many lives, despite the red coral and the bone from the deer's heart!

It was now late in 1503; Nicolaus Copernicus had been away from Poland ever since 1491, apart from his one brief visit to be installed as a canon of Frombork, and the time had come for him to return home. So he left Italy—not without regrets, one may well imagine; he had many friends there, and he had received the best possible education. At the age of thirty, his travels abroad were over. He went back to Warmia, and joined his uncle, Bishop Łukasz, whose episcopal seat was at the quiet town of Lidzbark. He never again left Poland.

Warmia (or Ermland) is in the northernmost part of Poland, and is low-lying, with cold winters. There are plains, pine forests and brightly-

Opposite: *Woodcut of an operation from Johannes de Ketham's* Fascicolo Medicina, *1493. Much of Copernicus' time was spent in medical studies*

coloured summer flowers, though for months at a time everything is mantled with snow. Lidzbark itself was dominated by its castle, and when Copernicus arrived there the castle was dominated by the Bishop. Łukasz Watzenrode had gained in power until he had become virtually the ruler of the whole of Warmia. We know that his personality was cold, and that he did not have the priceless gift of getting on with people, but from all accounts he was honest and high-principled. He was also firm, and the Teutonic Knights, who remained a constant menace, did not like him at all; the Grand Master of the Order once described him as 'the devil incarnate'. Łukasz was the trusted friend and adviser of three kings in succession: John Albert, Alexander (not to be confused with the poisoning Pope) and Sigismund; and his influence greatly strengthened the ties between Warmia and Poland proper.

We know, too, that he was a patron of the arts, and that he took himself very seriously. Pomp and ceremony were the rule. In Lidzbark, a gun

In 1503 Copernicus joined his uncle, Bishop Łukasz, at Lidzbark Castle in Warmia

was fired to indicate that the Bishop was sitting down to have dinner; in Frombork, he ordered that the canons should carry his bishop's staff ahead of him as he walked along – and when they objected, he said that at least he would make his nephews do so (though we do not know whether this actually happened).

Life at Lidzbark Castle was lively and varied. There were crowds of visitors, ranging from local knights to foreign emissaries, and in many ways the castle was not unlike a feudal court. The total number in the Bishop's household was, apparently, over fourteen hundred. Nicolaus Copernicus lived in the castle, and we know that he made some astronomical observations from there; he recorded an eclipse of the Moon in 1511, and also made some notes about the movements of the planet Mars. But officially Copernicus was the Bishop's secretary and physician, and his life was a busy one.

Łukasz did not spare himself. There were constant journeys; political and royal ceremonies, diplomatic conferences, conventions, meetings, negotiations, Church functions – the Bishop attended them all, and Copernicus was generally by his side. His travels took him all over Warmia, and he renewed his acquaintance with his old home, Toruń, as well as with the busy port of Gdańsk. He was no mere 'yes-man'. His advice was not only sought, but also followed, and as time went by he made more and more decisions on his own account. The man whom we remember today as a quiet pioneer of astronomy was very much a political figure in the Warmia of Bishop Łukasz' reign.

It might be thought that this busy life would draw Copernicus away from any thoughts about the plan of the universe. Yet nothing could be further from the truth. He had made up his mind to 'study the stars', and nothing could divert him. And it was at Lidzbark, while he was so active both politically and socially, that he wrote the first book in which he claimed, categorically, that the Ptolemaic theory must be wrong.

It was a short book, less than twenty pages long, and it was not printed. The title is *Nicolai Copernici de Hypothesibus Motuum Cælestium a se Constituis Commentariolus*, but it is always known simply as the *Commentariolus* (or *Commentary*). Handwritten copies were circulated among Copernicus' scientific friends – chiefly those in the Cracow area – and only three copies survive today: one in Vienna, one in Stockholm and one in Aberdeen, the latter being a copy by one Duncan Liddell, who lived in the second half of the sixteenth century. No printed versions were available for more than three hundred and fifty years after the original manuscript was written, which must have been about 1507. We do know that a Cracow professor, Maciej Miechowita, was in possession of a copy in 1514, and by then it was reasonably well-known among Polish scholars, so that it must have been in circulation for some time. (Tycho Brahe, the second great figure in the story of the scientific revolution, received a copy of it in 1575.)

It would be difficult to over-estimate the importance of the *Commentariolus*. It is essentially an outline sketch, without any mathematical reasoning; and it does not represent a sudden leap from the old, incorrect system to a new, accurate one. In fact Copernicus never did produce a satisfactory world system, mainly because he was convinced that all celestial orbits must be perfectly circular, and this forced him to keep to the epicycle-and-deferent idea. But he did realize that by taking the Earth away from the centre of the system, and putting the Sun there instead, he could make things very much simpler and more logical. To quote from the *Commentariolus*:

'Whatever motion we may observe in the firmament does not originate from the firmament itself, but from the motion of the Earth. The Earth, consequently, with its nearest elements undergoes a revolution over a period of twenty-four hours in its unchangeable poles, and the firmament, together with the highest heavens, remains motionless. . . . It would seem most unreasonable to attribute motion to what encompasses and affords space, rather than to that which is encompassed and contained, as is the case with the Earth.'

There are seven basic assumptions in the *Commentariolus*. I quote them here from the translation made from the two oldest-known copies, the Vienna and the Stockholm. They are:

1   All the celestial circles or spheres do not have just one [common] centre.
2   The centre of the Earth is not the centre of the universe, but only of gravity and of the lunar orbit.
3   All the spheres revolve round the Sun, as if it were in the middle of everything, so that the centre of the world is near the Sun.
4   The ratio of the Earth's distance from the Sun compared with the height of the firmament is so much smaller than the ratio of the Earth's semi-diameter to the distance from the Sun that the distance to the Sun is insignificant when compared with the height of the firmament.
5   The motions appearing in the firmament are not its motions, but those of the Earth. The Earth with its adjacent elements (i.e. air and water) performs a daily rotation around its fixed poles while the firmament remains immobile as the highest heaven.
6   The motions of the Sun are not its motions, but the motion of the

Earth and our sphere with which we revolve around the Sun just as any other planet does; so the Earth is carried along by several motions.

7 What appears to us as retrograde and forward motion of the planets is not their own, but that of the Earth. The Earth's motion alone, therefore, is sufficient explanation for many different phenomena in the heavens.

And he also gives the following sentence, showing that he knew the order in which the planets move round the Sun:

'The highest is the sphere of the fixed stars, containing and fixing location for everything. Below it is Saturn, followed by Jupiter, then Mars; below it the sphere in which we move, then Venus and finally Mercury. The lunar sphere revolves around the centre of the Earth.'

The essential step had been taken. Not, of course, that Copernicus had solved everything at a single stroke. He had produced a pattern which was on the right lines – but he still needed his epicycles, and he ends the *Commentariolus* by saying that it needs thirty-four circles to explain 'the structure of the world and the entire dance of the planets'.

Most people assume that the start of the great revolution was the year of 1543, with the publication of Copernicus' main book. In fact, there are grounds for claiming that it started with the *Commentariolus* over thirty years earlier; but the restricted circulation meant that comparatively few people outside Poland knew about it – though the Papal Court in the Vatican had certainly heard of 'Copernicanism' by 1533, and two years later Bernard Wapowski, the map-maker who had known Copernicus for so long, was discussing it quite openly. For the moment there was no adverse criticism. No doubt the Church authorities in Rome felt that here was yet another vague theory which would soon die.

Then, in early 1512, there came a tragedy which altered the whole course of Nicolaus Copernicus' life. Łukasz Watzenrode had been to Cracow for the wedding of King Sigismund and the coronation of the young Queen, Barbara Zapolya. His nephew had not accompanied him, but had remained in Lidzbark Castle, no doubt to deal with official matters. It was bitterly cold; snow covered the ground, and on the way home the Bishop was taken ill. He had to be taken to the nearest place of shelter, which happened to be Toruń, but the illness developed quickly. A message was sent to Lidzbark; Nicolaus set out, but by the time he reached Toruń he found that Łukasz Watzenrode had died. He was only 65 years old, but his hard-working life had at last taken toll of his energies. His body was taken to Frombork, where, with all due honours, he was buried in the cathedral grounds.

Copernicus' career at Lidzbark was over. He left the castle, and went

65

to Frombork; he was a canon of Warmia, and Frombork was the seat of the Chapter. He took up residence in the north-west tower which adjoins the ancient cathedral, and it was here that he spent most of the rest of his life. 'In remotissimo angulo terræ' . . . The remotest corner of the Earth. This was how Copernicus described it, but it was his home now, and it was at Frombork that he carried out the greatest of all his studies.

Though Nicolaus Copernicus had been to Frombork only sporadically, he was well-known there by the time that he arrived to make his home in the cathedral house. In particular, Andrzej was already installed as a canon, and this may be the moment to follow through his story, which is in such stark contrast to that of his brother. Andrzej had picked up some disfiguring disease, possibly while in Italy. It may well have been leprosy. Nicolaus tried all the medical skill that he had learned in Padua; Andrzej went off on a year's leave of absence to search for a cure, but all in vain. He returned to Frombork, and grew slowly but steadily worse. And then, in 1512, he was driven out in case he infected his companions. It does not sound a very Christian act on behalf of the Chapter; but it happened, and Andrzej left. We do not know what became of him, but it seems certain that he died in or before 1519, probably in Rome.

No doubt Nicolaus was distressed. Equally certainly, he was powerless to help. We do not know the closeness of the ties between the brothers, though since they had spent so many of their student days together we may assume that they were considerable. But now Andrzej was gone for good, and Nicolaus was left with no near relatives to hand, though there were plenty of distant ones. Note, too, that he never married, and in the conventional sense there never seems to have been 'a woman in his life', though we will come later to the depressing story of Anna Szylling and the overbearing Bishop Dantyszek.

Soon after reaching Frombork, Copernicus built a wooden quadrant for measuring the position of the Sun in the sky. Another instrument which he used, and presumably made, was a triquetrum, used for checking the positions of the stars. Unfortunately, neither these nor any other of Copernicus' astronomical instruments have come down to us; we know that at one stage the triquetrum was in the possession of Tycho Brahe, but it has long since been lost. Not, of course, that Copernicus himself was a particularly skilful observer, and his results

Opposite : The courtyard of the d'Este Palace in Ferrara. Here Copernicus received the conferment of his Doctorate in Canon Law in 1503

were appreciably less accurate than those which the Greeks had made so long before. He was a theorist, and his observational work was very limited.

It has often been said that Copernicus never managed to see the planet Mercury. It is true that Mercury, the smallest and innermost of the planets, is never very conspicuous; with the naked eye, it is visible only when low in the west after sunset or low in the east before sunrise, and it can never be seen against a dark background, so that modern studies of it have to be made when both it and the Sun are high in the sky. This involves using a telescope with setting circles, and in Copernicus' time there were no telescopes at all. On the other hand, the legend that he never saw Mercury because of mists rising from the Vistula seems to be quite out of court. When I went to Toruń for the Copernicus quincentenary celebrations in 1973, I made a careful check on the stars low over the dusk horizon. I could see stars much fainter than Mercury, even when very low down; and presumably the atmosphere of Toruń in the early sixteenth century was much less polluted and dust-laden than it is today, quite apart from the interference by modern electric lights. Moreover, remember that Copernicus spent over ten years under the clear skies of Italy. All in all, the Mercury tale seems to belong to the same class as that of Canute and the waves—or, for that matter, Galileo and the weights dropped off the Leaning Tower of Pisa.

Be this as it may, Copernicus certainly made a number of observations. And from 1515 or thereabouts he was working on the manuscript of his great book; though publication was deferred for a variety of reasons, the book was probably complete by 1533. But for the first part of his period at Frombork he was faced with the perennial problem of the Teutonic Knights, and he had to become a military organizer as well as a physician and a statesman.

The Knights were becoming increasingly active. Armed gangs raided Warmia, and there were endless 'border incidents'. Copernicus did his best, and took part in negotiations with the Grand Master of the Order; so did the new Bishop of Warmia, Fabian, who was admittedly somewhat weak and vacillating. Yet things went from bad to worse, and the storm finally broke on New Year's Day 1520, with open war. The Knights occupied the town of Braniewo, and their leader, who rejoiced in the typically Germanic name of Albrecht Hohenzollern, proclaimed himself the legal defender of Warmia against the Poles, claiming that he had the backing of the Pope (which was probably true). Copernicus went to Braniewo, this time in his official capacity as the Bishop's deputy. Negotiations were useless; the Knights advanced, and by the end of

January they had reached Frombork, devastating everything outside the fortified walls.

Copernicus went to Olsztyn, where there was an extremely well-built castle. He stayed there for three years, from 1516 to 1519, and became the mainstay of resistance to the marauding Knights. The Olsztyn garrison consisted of a few hundred Polish soldiers together with local help, and most of the canons left hurriedly for the safer refuge of Gdańsk. Copernicus was made of sterner stuff, and he reinforced the fortifications as well as laying in supplies of food and arms. Eventually, royal troops arrived to help, and the Knights were compelled to raise the siege. They also failed to take Lidzbark Castle, and in the spring of 1521

Celestial Globe of Martin Bylica of Olkusz, made by Hans Dorn in 1480, and used by the students at Cracow University

a four-year cease-fire began, though armed raids continued almost unabated. Copernicus, now in virtually supreme charge of Olsztyn, undertook relief operations, and it is fair to say that his skill and courage played a great rôle in the whole of the war.

In the end the Knights had to admit failure. On 10 April 1525 the Grand Master rendered homage to King Sigismund I in the market-place of Cracow, and the main danger was at an end. Though the Teutonic Knights were still in existence as an organized body, they never again attempted full-scale invasion, and their power dwindled steadily. The Order was finally suppressed in the early years of the nineteenth century, by Napoleon Bonaparte. Few people regretted its passing.

It is quite remarkable that during his two spells at Olsztyn (1516–19 and 1520–21) Copernicus still found time to carry on some astronomical observation, but he had to suspend work on his book, and did not begin again seriously until after the war had ended and he had returned to Frombork – this time for good. Even then he was still heavily occupied with administrative duties, and, of course, with his work as a physician to the people, from dukes to peasants. This is no place to delve deeply into some of his other activities; it will be enough to mention that he was concerned with the reform of the coinage of the country, and was the first to state a well-known principle that bad or debased money will drive out good. (This is known as Gresham's Law, but priority must certainly go to Copernicus.) He had been invited to take part in a conference about reforming the calendar, but declined on the grounds that the movements of the Sun and Moon were not sufficiently well known; in his letter of rejection he added, significantly, that he was working on the problem. He had a great sense of responsibility, and there is no reason to doubt that he was absolutely honest and straightforward in his dealings with his subordinates as well as his superiors – something which was rare in the sixteenth century as it is today.

The defeat of the Teutonic Knights may be said to mark the end of the adventurous part of Copernicus' career. Thereafter he stayed at Frombork, travelling only over the local area which was his particular responsibility in order to deal with the affairs of the people; his was a life of ordered routine, and it suited him well. Now, at long last, he could devote his main mental energy to the completion of the theory which, he was sure, would cause a complete revision of Man's ideas about the universe in which we live.

'De Revolutionibus . . .'

Nobody can be quite sure when Copernicus' great book was finished. It may have been complete by 1530, and certainly by 1533. Yet for some time it remained unpublished, and on the whole it was fortunate to have been rescued. Had it been lost—as it could well have been—then who today would remember Nicolaus Copernicus? The answer is: 'Nobody', apart from a few historians concerned with the general story of the Teutonic Knights and their raids. The *Commentariolus,* as we have seen, did not come to light until much later, and its circulation was very restricted. So we must admit that the whole of Copernicus' reputation depends upon one book: *De Revolutionibus.*

Copernicus wrote out the manuscript in his neat longhand. The manuscript now stored in the Jagiellonian Library in Cracow runs to over two hundred pages, together with diagrams, and is subdivided into six sections, of which the first deals with the arrangement of the Solar System; the second, with a newly-arranged star catalogue; the third, with precession; the fourth, with the movements of the Moon; and the last two, with the movements of the planets. Obviously, the cornerstone of the whole Copernican edifice is the claim that the Sun, not the Earth, lies in the centre of the system, and it is this which makes the book immortal. (I repeat that the *Commentariolus* was merely a preliminary essay.) Yet there is one distinction which must be made between the Copernican 'theory', which was right, and the Copernican 'system', which was just as wrong as Ptolemy's had been. According to his 'system', all celestial orbits are perfectly circular, and so there was no alternative but to come back to the clumsy idea of epicycles, deferents and similar contrivances. It is also true to say that Copernicus was quite unable to give any firm scientific proof of his theory; and because of the faults in the system, it provided results which were only marginally better than those based on the Ptolemaic universe. In other words, he

71

was open to criticism on scientific as well as religious grounds, as he may well have realized. Remember, too, that his astronomical instruments were not particularly accurate, and that Copernicus himself made no pretence of being a skilful observer.

Yet in spite of all these deficiencies, *De Revolutionibus* was certainly the most important astronomical book ever written up to that time. I do not propose to summarize it here, but I have selected a few quotes in order to show both the way in which Copernicus' mind worked, and the calm, unemotional manner in which he put forward his views; there is no trace of conceit or arrogance—a remarkable contrast with Tycho Brahe and Galileo, as we will see later! Here, then, are some passages from the book.

'First of all we assert that the universe is spherical; partly because this form, being a complete whole, needing no joints, is the most perfect of all; partly because it constitutes the most spacious form, which is thus best suited to contain and retain all things; or also because all discrete parts of the world, I mean the Sun, the Moon and the planets, appear as spheres; or because all things tend to assume the spherical shape, a fact which appears in a drop of water and in other fluid bodies when they seek of their own accord to limit themselves. Therefore no-one will doubt that this form is natural for the heavenly bodies.'

'That the Earth is likewise spherical is beyond doubt, because it presses from all sides to its centre. Although a perfect sphere is not immediately recognized because of the great height of the mountains and the depression of the valleys, yet this in no wise invalidates the general spherical form of the Earth. This becomes clear in the following manner: to people who travel from any place to the north, the north pole of the daily revolution rises gradually, while the south pole sinks a like amount. Most of the stars in the neighbourhood of the Great Bear appear not to set, and in the south some stars appear no longer to rise. Thus Italy does not see Canopus, which is visible to the Egyptians.'

(By means of its motion) 'the Sun measures for us the year, the Moon the month, as the most common units of time. And thus each of the other five planets completes its orbit. Yet they are peculiar in many ways. First, in that they do not revolve about the same poles around which the first motion takes place, progressing instead in the oblique path of

Opposite: *Part of the fortifications of Frombork Castle.*

the Zodiac; second, in that they do not seem to move uniformly in their own orbits, for the Sun and the Moon are discovered moving now with a slower, now a faster motion. The remaining five planets, moreover, we also see at times going backward and, in the transition, standing still. And while the Sun moves along always in its direct path, the planets wander in various ways, roaming now to the south, now to the north. Wherefore they are designated "planets". They have the added peculiarity that they at times come nearer to the Earth, where they are called "at perigee", and then again recede from it, where they are called "at apogee". Nevertheless, it must be admitted that the motions are circular, or are built up of many circles; for thus such irregularities would occur according to a reliable law and a fixed period, which could not be the case if they were not circular.'

'Since it has already been proved that the Earth has the shape of a sphere, I insist that we must investigate whether from its form can be deduced a motion, and what place the Earth occupies in the universe. Without this knowledge no certain computation can be made for the phenomena occurring in the heavens. To be sure, the great majority of writers agrees that the Earth is at rest in the centre of the universe, so that they consider it unbelievable and even ridiculous to suppose the contrary. Yet when one weighs the matter carefully, it will be seen that this question is not yet disposed of, and for that reason is by no means to be considered unimportant. . . . Now the Earth is the place from which we observe the revolution of the heavens and where it is displayed to our eyes. Therefore, if the Earth should possess any motion, the latter would be noticeable in everything that is situated outside of it, but in the opposite direction, just as if everything were travelling past the Earth. And of this nature is, above all, the daily revolution. For this motion seems to embrace the whole world, in fact everything that is outside of the Earth, with the single exception of the Earth itself. But if one should admit that the heavens possess none of this motion, but that the Earth rotates from west to east; and if one should consider this seriously with respect to the seeming rising and setting of the Sun, of the Moon and the stars; then one would find that this is actually true.'

'Now, whether the world is finite or infinite we will leave to the quarrels of the natural philosophers; for us remains the certainty that the Earth, contained between poles, is bounded by a spherical surface. Why should we hesitate to grant it a motion, natural and corresponding

Opposite: *Olsztyn Castle, Poland.*

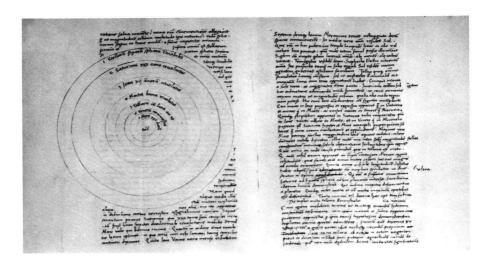

Pages from the manuscript of De Revolutionibus *with Copernicus' drawing of the heliocentric system. The manuscript was completed in 1533, but was not published until 1543*

to its form, rather than assume that the whole world, whose boundary is not known and cannot be known, moves? And why are we not willing to acknowledge that the *appearance* of a daily revolution belongs to the heavens, its *actuality* to the Earth? The relation is similar to that of which Virgil's Æneas says: "We sail out of the harbour, and the countries and cities recede." '

'Since nothing stands in the way of the mobility of the Earth, I believe we must now investigate whether it also has several motions, so that it can be considered one of the planets. That it is not the centre of all the revolutions is proved by the irregular motions of the planets, and their varying distances from the Earth, which cannot be explained as concentric circles with the Earth at the centre. . . . If one admits the motionlessness of the Sun, and transfers the annual revolution from the Sun to the Earth, there would result, in the same manner as actually observed, the rising and setting of the constellations and the fixed stars, by means of which they become morning and evening stars; and it will thus become apparent that also the haltings and the backward and forward motion of the planets are not the motions of these but of the Earth, which lends them the appearance of being actual planetary motions. Finally, one will be convinced that the Sun itself occupies the centre of the universe.'

One more extract—this time from the most important section of Book One of *De Revolutionibus*, which sums up the whole picture:

'I also say that the Sun remains forever immobile, and that whatever apparent movement belongs to it can be verified as due to the mobility of the Earth that the magnitude of the world is such that although the distance from the Sun to the Earth in relation to the orbital circles of the planets possesses magnitude which is sufficiently manifest in proportion to those dimensions, this distance, as compared with the sphere of the fixed stars, is imperceptible. . . . In the deduction of terrestrial movement, we will however give the cause why there are phenomena such as to make people believe that even the sphere of the fixed stars somehow moves. Saturn, the first of the wandering stars, follows; it completes its circuit in 30 years. After it comes Jupiter, moving in a 12-year period. Then Mars, which completes a revolution every two years. The place fourth in order is occupied by the annual revolution in which the Earth is contained, together with the orbital circle of the Moon as an epicycle. In the fifth place, Venus, which moves around in nine months. The sixth and final place is occupied by Mercury, which completes its revolution in a period of 80 days. In the centre of all rests the Sun. For who would place this lamp of a very beautiful temple in another or better place than this, wherefrom it can illuminate everything at the same time? . . . Now the careful observer can see why progression and retrograding appear greater in Jupiter than in Saturn and less than in Mars; and in turn greater in Venus than in Mercury; in addition, why when Saturn, Jupiter and Mars are in opposition they are nearer to the Earth than at the time of their conjunction and their reappearance, and especially why at the times when Mars is in opposition to the Sun it seems to equal Jupiter in brilliance and to be distinguished from it only by its reddish colour, but when discovered through careful observation at other times by means of a sextant is found with difficulty among the stars of the second magnitude. All these things proceed from the same cause, which resides in the movement of the Earth.'

I do not apologize for giving these rather lengthy quotes, for the excellent reason that Copernicus describes his theory better than anyone else could do. There are, however, a couple of points to be cleared up. In the language of the fifteenth and sixteenth centuries, the term 'world' has a variety of meanings, and is often taken to be synonymous with 'universe' rather than applying specifically to the Earth. Secondly, let me explain the opposition and retrograding of the planets, which I

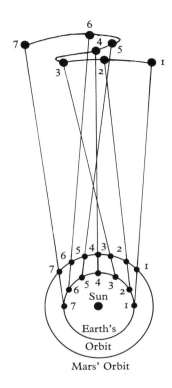

Figure 5: Retrograding of planets.

can do best by means of a diagram (Fig 5). A planet is in opposition when it is on the far side of the Earth with respect to the Sun, so that it is opposite to the Sun in the sky. Ordinarily, a planet will move against the stars from west to east; but around opposition the Earth is 'catching up' the planet and passing it, so that the planet seems to backtrack temporarily from east to west. This, of course, applies only to the superior planets (i.e. Mars, Jupiter, Saturn and the planets discovered since Copernicus' time). Venus and Mercury behave in rather a different manner, and obviously they can never come to opposition.

In various passages in the book, Copernicus disposes of the age-old objections to the idea of a rotating and moving Earth. He writes, for instance, that as the Earth spins round it takes its atmosphere with it, so that there is no fear of a constant howling gale. There is also the question of what is termed parallax–the apparent shift in position of a nearby object, against a more distant background, when the observer's position changes. The stars show no daily parallaxes which can be measured, and to Copernicus this was an extra proof that the stars are very remote indeed.

Such, then, are some of the main points made in the great book. Even while it lay in Copernicus' room in Frombork, being constantly examined and revised, hints of the new theory of the universe were being spread around, due mainly to the small but significant circulation of the *Commentariolus* and also the activities of some of Copernicus' friends, notably the map-maker Wapowski and Bishop Tideman Giese. In 1533 the Pope (Clement VII) asked his secretary to explain these strange theories about a central Sun. Then, in 1536, Nicolaus Schönberg, Cardinal of Capua, wrote to Copernicus asking for some extra explanations and tables–which he duly received, and which made him a firm convert. He even volunteered to pay all the expenses of bringing out the

Bishop Tideman Giese, a loyal friend of Copernicus

book and giving it to the world, but unfortunately he died shortly afterwards.

In the dedication of the book when it eventually appeared, Copernicus paid great tribute to Cardinal Schönberg and to Bishop Giese. He said that Giese had 'often encouraged me, and not infrequently with bitter reproaches, to publish this work, which had lain concealed in my abode, not only for nine years but even for four times nine years, and to let it finally come out into the light of day'. But perhaps the final impetus was given by Georg Joachin von Lauchen, better known as Rhæticus (because he was born in the province of Rhaetio, and followed the usual custom of Latinizing his name).

Rhæticus, a 25-year-old professor of mathematics at Wittenberg,

Frontispiece of Martin Luther's German edition of The Bible, *published in Wittenburg, 1536*

had heard about the heliocentric theory, and it appealed to him. As he wanted to know more about it, he decided to go to see Copernicus in person, and he arrived at Frombork in May 1539. He was, in fact, running something of a risk, because this was the time of the controversy centred round the religious reformer Martin Luther; the Catholic church of Warmia was intensely anti-Lutheran, and anyone from Wittenberg was automatically suspected of being a personal representative of Luther himself. Copernicus, of course, was entirely free from this sort of prejudice, and he received Rhæticus warmly, so that instead of paying a fleeting visit to Frombork the young professor stayed for two years.

By now Copernicus was in his mid-sixties, and had been in his 'remotest corner' for many years. He still kept up his administrative duties, and he was still famous as a doctor. (It is interesting to note that only two years before his death he was summoned by no less a person than Duke Albert of Prussia, the former Grand Master of the Teutonic Knights, to attend to a trusted friend who had been taken ill and whom the Prussian doctors seemed unable to cure. Copernicus went, and apparently his treatment was successful.) But his existence was rather a lonely one, and his first great biographer, the French astronomer Pierre

The Wittenburg Reformers, by Lucas Cranach circa 1530. Luther, left of *John Frederick of Saxony* centre, *Luther's followers on the* right *are Zwingli and Melancthon*

Gassendi, wrote that at that time 'he felt revulsion toward any form of familiarity and any frivolous and pointless conversation', so that no doubt he was glad to have Rhæticus with him.

Rhæticus made up his mind to press for the publication of the book. His first step was to write one of his own—a brief, 38 page volume which he called *Narratio prima* (*First Account*), and in which he outlined the Copernican theory in clear, concise prose. It was printed in Gdańsk, and was soon sold out, after which a new edition was produced in Basle.

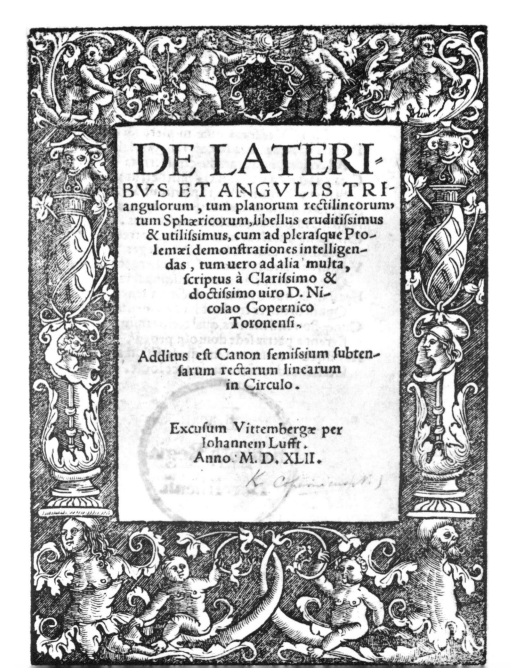

Rhæticus promised to follow it with a Second Account, dealing with other aspects of the theory, but there was no need for him to do so. At last Copernicus gave way to the pleadings of his friends, Giese and Rhæticus above all. The book could go forward.

Apparently Giese took the precious manuscript and passed it to Rhæticus, who in turn took it to a Nürnberg publisher, Petrejus, who was anxious to produce it. Rhæticus meant to see it through the press, but by ill-fortune he left Wittenberg in the autumn of 1541; he had been appointed to a new post in Leipzig, and he handed over the supervision to a Lutheran clergyman named Osiander. The choice was not a good one. Osiander was worried about possible attacks from the Church, and he suggested that it would be better if Copernicus wrote a preface stating that the theory was not meant to be taken literally. He actually wrote to Copernicus along these lines in 1541. Not surprisingly, he met with a rebuff–although the great astronomer did dedicate the book to the new Pope, Paul III, who was known to be a scholar and a lover of science. 'I am fully aware, Holy Father,' wrote Copernicus, 'that as soon as they hear in these volumes of mine about the revolutions of the spheres of the universe and that I attribute some sort of motion to the Earth, some persons will immediately raise a cry of condemnation against me and my theories.'

In fact the attacks had already started, and the first one of real importance came from Luther. Since Luther was convinced that everything in the Bible must be literally true, he had no patience with Copernicus, and he wrote: 'Mention has been made of some new astrologer, who wants to teach that the Earth moves around, not the firmament or heavens, the Sun and Moon. Just as though a man were to sit in a cart or a moving boat, and thinks he sits motionless and rests while the earth and the trees move. . . . This fool seeks to overturn the whole art of astronomy. But as the Holy Scriptures show, Jehovah ordered the Sun, not the Earth, to stand still.'

One can hardly admire either Luther's manners or his scientific knowledge. His onslaught was made in 1539, and another, two years later, came from his colleague Melancthon: 'Some people are of the opinion that it is excellent and clever to work out something as absurd as did this Sarmatian astronomer, who moved the Earth and stopped the Sun. Indeed, our wise rulers would do well to restrain such clever frivolity!' Calvin was another critic; and we must mention a hack playwright named Gnapheus, who produced a comedy called *The Wise Fool* in which he ridiculed Copernicus and his ideas. Note, however, that all these attacks were made on religious grounds. Before 1543 the main theory had not been published, so that all the critics had before them was the brief summary in Rhæticus' *Narratio prima* (disregarding the *Commentariolus*, which was not generally available).

Not surprisingly, Copernicus was unhappy about these signs of hostility. He had other troubles to face also. The new Bishop of Warmia was one Dantyszek, who had known Copernicus for many years—in fact, ever since the visit to Rome—and who had not been exactly a pillar of virtue. Copernicus had supported his candidature for the Bishopric, but when Dantyszek had been installed he became a moralist in the worst sense of the term. At that time Copernicus' housekeeper was one Anna Szylling, who may have been a distant relative of the Kopernik family. Anna was, we gather, a handsome young woman, and the Bishop decided that her presence was undesirable, so that he ordered her to be sent away. The miserable affair began in 1538, and dragged on for more than a year before Copernicus finally yielded; Anna left his house, and subsequently departed from Frombork altogether. Undoubtedly it was this which made the old astronomer's last years much more gloomy than they would otherwise have been, and one can have no admiration for Dantyszek, though it is only fair to add that most of his contemporaries regarded him as an enlightened scholar as well as a suave diplomat.

The Anna Szylling affair might have broken the spirit of a lesser man; but old and infirm though Copernicus had become, his mind was as active as ever, and he was now determined that his book should be published. Printing was started under the watchful eye of Osiander, and on or about 21 March 1543 the great work was ready in Nürnberg. We know it as *De Revolutionibus Orbium Cælestium*, which may be translated as *On the Revolutions of the Celestial Orbs*, though it may be that Copernicus himself called it simply *De Revolutionibus*. Also, there was a preface in which it was stated that the text was presented as a pure theory, not as literal truth. The preface was written by Osiander, but since it was unsigned many people jumped to the conclusion that it was the work of Copernicus himself. Tideman Giese, for one, was furious, and even sent a formal protest to the Nürnberg Senate.

After this lapse of time it is difficult to be sure of Osiander's motives. Quite possibly he was anxious to soften any criticism which might be levelled at the book on religious grounds, and if this is true then we must admit that he succeeded, even if only for a while. Initially the book made surprisingly little impact, and the real trouble did not begin for some years afterwards. But Copernicus himself was no longer there; he died on 24 May 1543. Legend tells that he was able to see the first copy of his book in his last moments, but this is probably wrong, because he had been unconscious for some days before his death. So he never saw *De Revolutionibus* in print—and neither did he ever know about Osiander's unauthorized preface.

For some time the Church authorities took no official notice of the Copernican theory, possibly because they regarded it as no more than a passing fad. News of it spread; in England, for instance, it was championed by Robert Recorde, who was an astrologer and mathematician as well as physician to Queen Mary Tudor. Others who became, apparently, 'Copernicans' were John Dee and Thomas Digges, two rather shadowy figures of Elizabethan England who may have accomplished more than we know. Yet the first tables prepared on the Copernican theory (Reinhold's *Prutenic Tables* of 1551) were very little better than the Ptolemaic ones had been—because, as I have stressed several times, Copernicus' system was wrong. Even he, with his courage and insight, could not bring himself to break free from the concept of perfectly circular orbits.

Copernicus was a great man as well as a great thinker. We know little to his detriment, and much to his credit. He may not have been a truly colourful character, but he rose to the occasion when need be; witness his temporary transformation from a physician-cum-scholar into a military organizer against the Teutonic Knights. His contribution to science was invaluable, and history will never forget the quiet man from 'the remotest corner of the Earth'.

Nicolaus Copernicus as an old man; from a late sixteenth century woodcut

The Master of Hven

It is interesting that of the first four great characters in the story of the scientific revolution, two (Copernicus and Kepler) were quiet and modest, while the others (Tycho Brahe and Galileo) were quite the reverse. Nobody could say that there was anything shy or retiring about either of them, and in each case they met trouble more than half-way. Let us begin, then, with Tycho, who was no theorist, but whose painstaking, amazingly accurate observations paved the way for those who followed him.

He was a Dane of noble ancestry; his father Otto was Governor of Helsingborg Castle, and a man of very considerable influence. Tyge, or Tycho, his eldest son and the second of the five children, first saw the light of day on 14 December 1546, at the family seat at Knudstrup in Skaane–then Danish, now included in Sweden. (To be accurate there were six children, but one of them–Tycho's twin brother–died either at birth or a few hours afterwards.) Early events set the pattern for the tempestuous life which was to follow. For some unknown reason Otto had promised his brother Jørgen, an officer in the Danish Navy, that as soon as he had another son it would be time to hand little Tycho over to be brought up in Jørgen's household. When another baby boy was born, Otto and his wife Beate apparently had second thoughts, with the result that Uncle Jørgen kidnapped Tycho and removed him. We do not know the full story, but it does seem certain that Tycho spent his youthful years at Torstrup, Jørgen's home, rather than with his true parents. It also seems that after a while the situation was amicably resolved, and there were no hard feelings on either side.

Actually we do not know much about the boy's upbringing, but

Opposite: Tycho Brahe's statue on the Island of Hven, Denmark, the site of his two observatories

probably it was conventional enough by the standards of the time, and the first step after school was University. So in 1559 Tycho went to Copenhagen (it was then, incidentally, that he started to sign himself 'Tycho' instead of the Danish 'Tyge'). During his stay there, he saw an eclipse of the Sun. The date was 21 August 1560; the eclipse was total in Portugal but only partial in Denmark, so that the glorious pearly corona could not be seen. Nonetheless, the sight made a lasting impression on Tycho's receptive mind, and it may be that his love of astronomy began then.

Star-gazing was all very well so far as Tycho was concerned, but it did not suit his uncle, who was firmly set upon making his nephew an important figure in Danish political circles. Clearly, then, law should be the chosen study, and Leipzig University had an excellent law school. However, Jørgen did not trust Tycho's dedication to legal studies, and he probably cast unfavourable looks at the Latin translation of Ptolemy's *Almagest* which the boy had purchased while still at Copenhagen

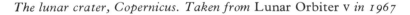

The lunar crater, Copernicus. Taken from Lunar Orbiter v *in 1967*

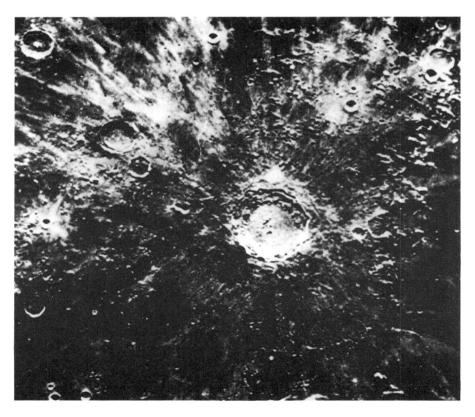

(doubtless with money taken from his allowance). So a travelling companion or guardian, Anders Vedel, was selected to go to Leipzig too, and, metaphorically, keep Tycho's nose to the legal grindstone. It must have been a difficult task. Vedel was only four years older than his charge, but he was intelligent and versatile, and he was too sensible to try to cut Tycho off from astronomy altogether. A bargain was struck; the student could look at the stars by night, provided that he put in a full day's work at law first. It must be added that Tycho very often carried out secret observations while Vedel was asleep, and for a time there was a certain chilliness between them, though it did not last for long and they eventually became close friends.

Then, in the year 1563, there came an event which made Tycho think very hard about the make-up of the universe. There was a conjunction of two bright planets, Jupiter and Saturn, which appeared

The lunar crater, Tycho. This view taken from Lunar Orbiter v, *looks down into the crater, showing the central peak, rough floor and precipitous walls*

side by side in the sky. Of course they were not really close together; Saturn is much more remote than Jupiter, but on 24 August the two seemed almost to merge into a single blob of light. Astronomical tables could be used to predict a conjunction of this sort, and Tycho realized that all the published information was faulty. The old *Alphonsine Tables*, worked out according to Ptolemy's theory, gave a date which was a full month wrong. Reinhold's *Prutenic Tables* of 1551, based on the Copernican system, were better, but there was still a discrepancy of several days.

This was Tycho's first recorded observation, and he made it in a decidedly elementary way. His only equipment consisted of a pair of compasses. By holding the centre close to his eye, and pointing one arm to Jupiter and the other to Saturn, he was able to find the angular distance between them by checking the compasses against a circle drawn on paper and divided into degrees. Primitive methods of this kind did not satisfy him, and he began to make proper measuring instruments, though he was still officially at Leipzig to study law and to consider astronomy as nothing more than a passing hobby.

In 1565, a letter arrived from Jørgen Brahe calling both young men home. Tycho was probably pleased; the more he studied law, the less he liked it. But war had broken out between Denmark and Sweden, and Jørgen was a vice-admiral in the Danish Navy, so that he was naturally preoccupied. Not long after his nephew's return, Jørgen died – in rather strange circumstances. King Frederik II of Denmark (later to play so important a rôle in Tycho's career) was crossing the bridge leading to Copenhagen Castle when he fell into the water. Jørgen pulled him out, but the soaking caused him to catch pneumonia, and nothing could be done to save him. Tycho was genuinely distressed; but there was nothing now to keep him at home or force him to continue with his law studies. On the advice of another uncle, Steen Bille, he decided to see to his affairs in Denmark (he had, of course, inherited some property from Jørgen) and then go back to Germany to study what really interested him.

He went first to Wittenberg, and then to Rostock. It was here that he met with the misadventure which has become so famous. For some reason or other he quarrelled with another Danish nobleman, Manderup Parbsjerg, during a party at the home of one of the professors, and a week later they took part in a duel in the best tradition of mediæval

Opposite: *Tycho Brahe and his great mural quadrant, and in the background various other instruments used at Uraniborg. In the basement, his alchemical laboratory. From Tycho's* Astronomicæ Instaurata Mechanica, *1587*

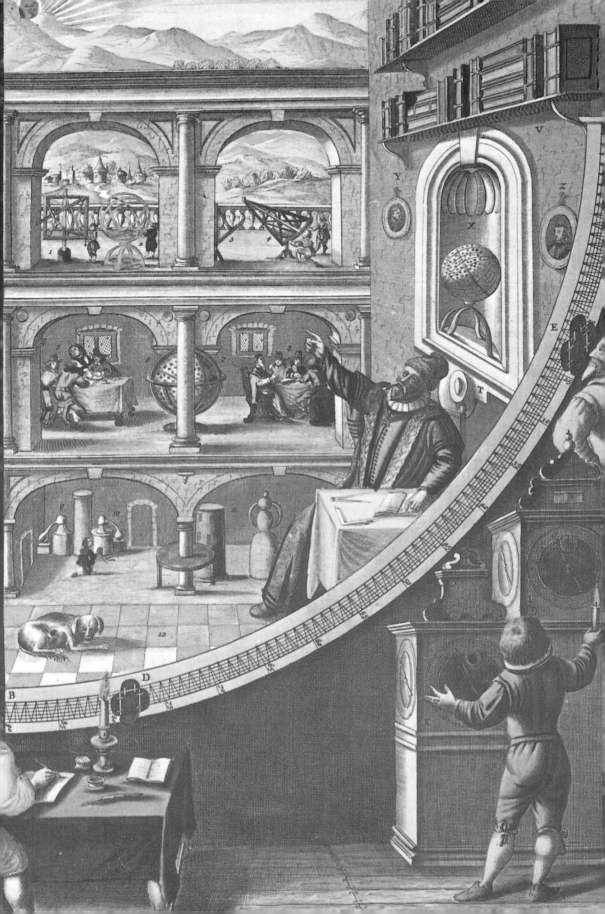

knights. Evidently Tycho had the worst of matters, and his nose was badly damaged. From all accounts the main injury was on the bridge, and the wound was disfiguring. Ever resourceful, Tycho carried out repairs with gold, silver and wax, and this was how his nose remained for the rest of his life. There is no record that it caused him any discomfort, and apparently he and Parbsjerg subsequently became good friends.

The duel took place just after the Christmas of 1566. Another incident which occurred during the Rostock period is worth recalling. Tycho was becoming more and more attracted to astronomy, but he was also well-disposed toward astrology, and he made the public statement that an eclipse of the Moon in October 1566 foretold the death of the Sultan of Turkey, Suleiman. Before long news came through that the Sultan had indeed departed this earthly life; but it later transpired that he had died before the eclipse, not afterwards, which cast some doubt upon the validity of the connection!

Another place of call during those formative years was Augsburg, where the first of Tycho's large observing instruments were set up, largely at the instigation of an alderman named Paul Hainzel. One was a large quadrant with a radius of nineteen feet, used for measuring the altitudes of the stars; the other was a sextant, for measuring the angular distances between stars (a distinct improvement on the pair of compasses which had been pressed into service some years earlier). The oaken quadrant proved to be very successful, and was actually used for making some measurements of the new star of 1572, about which more will be said later. Unfortunately the quadrant was destroyed in a storm in 1574, and we have also lost an ornate five-foot star-globe, which was sent from Augsburg to Hven, Prague and then Copenhagen before it was burned in the great fire of 1728.

In 1571 Tycho was back in Denmark, not for pleasure but because his father Otto was dying. After the funeral he stayed on for a while, and seems to have taken more than a passing interest in chemistry, but then came the turning-point of his career: the appearance of what we call the new star of 1572, now known to have been a colossal stellar explosion of the type we know as a supernova.

Let us hear the story in Tycho's own words.

'In the evening, after sunset, when, according to my habit, I was contemplating the stars in a clear sky, I noticed that a new and unusual

Opposite: *Tycho Brahe, 1546–1601 : anonymous portrait*

star, surpassing all the other stars in brilliancy, was shining almost directly above my head; and since I had, almost from boyhood, known all the stars of the heavens perfectly (there is no great difficulty in attaining that knowledge), it was quite evident to me that there had never

Tycho Brahe. This engraving clearly shows the part of his nose, cut off in a duel, which Tycho remade with gold, silver and wax

94

before been any star in that place in the sky, even the smallest, to say nothing of a star so conspicuously bright as this. I was so astonished at this sight that I was not ashamed to doubt the trustworthiness of my own eyes. But when I observed that others, too, on having the place pointed out to them, could see that there really was a star there, I had no further doubts. A miracle indeed, either the greatest of all that have occurred in the whole range of nature since the beginning of the world, or one certainly that is to be classed with those attested by the Holy Oracles.'

What made this so important? Simply the fact that according to Aristotle and the other ancient teachers, the heavens and the starry sphere were changeless. Yet this new star was more than normally obtrusive; it could not be overlooked, and next day Tycho found that he could even see it with the Sun well above the horizon. It was, indeed, as bright as the planet Venus. Tycho was not the first to discover it. From all accounts, priority must go to Wolfgang Schuler of Wittenberg, who saw it on 6 November. Next night it was seen by Hainzel at Augsburg and also by B. Lindauer at Winterthur in Switzerland. Tycho's original observation of it was not made until 11 November; but because of his close studies of it, it is always remembered as Tycho's Star.

Again I must digress, because the story of the star will be incomplete without an explanation of how unusual it really was. Stars—including the Sun—are not burning. Instead, they are creating their energy by what are called nuclear transformations inside them. To take the Sun as an example: near its core, the temperature is tremendous (of the order of 14 million degrees Centigrade, 25 million degrees Fahrenheit) and the pressure is colossal, so that very odd things are happening to the atoms of hydrogen, which make up a large proportion of the Sun's material. Basically, the hydrogen atoms—or, to be more precise, the hydrogen nuclei—are banding together to form atoms of another substance, helium. Each time the process takes place, a little energy is set free and a little mass is lost. It is this energy which keeps the Sun shining; and the mass-loss amounts to four million tons every second, though the Sun is so huge that it will not alter much for several thousands of millions of years yet.

Stars of solar type are in a stable condition, and shine steadily over very long periods (which is fortunate for us). However, toward the end of its brilliant career a star may suffer outbursts, producing what we call novæ. In an ordinary nova, the outburst affects only the outer layers, so that after a few months or a few years the star returns to its old state, apparently none the worse. But with a very massive star which has used up its main reserves of energy, there may be a catastrophic collapse, with intense heating of the core, so that the star literally blows much of its material away into space in what is termed a supernova explosion. For a short time the output of energy is staggering, but when the outburst is spent all that is left is a patch of expanding gas together with a very small, super-dense 'stellar wreck' made up of particles called neutrons. In fact, a supernova represents the death-agony of a very massive star.

The Sun will never become a supernova; it is not heavy enough, and in the course of time it will come to a quieter end. And not many super-novæ have been seen in our Galaxy in recorded times (though we have observed plenty in the outer galaxies; because a supernova is so power-ful, it can be seen over a vast range). The only galactic supernovæ about which we have any definite information are those of 1006, 1054, 1572 and 1604. The first was in the southern constellation of Lupus, the Wolf, and we know little about it. The 1054 supernova, in Taurus, was seen by Chinese and Japanese astronomers, and has proved to be of the utmost importance, because we can still see its remains; it was the forerunner of the gas-patch called the Crab Nebula, which sends us radio waves as well as x-rays and visible light, and which contains the only neutron star which has been optically identified. (Other neutron stars are traceable because of their rapidly-varying radio signals; we know them as pulsars). The 1572 supernova was Tycho's Star, and that of 1604 is always associated with the name of Kepler, so that discussion of it can be deferred for the moment (Fig 6).

Tycho, needless to say, had no idea of the true nature of the spectacular newcomer, but at least he could make precise measurements of it—and he did. Using the equipment which he had devised, he checked on the position, near the conspicuous 'W' of stars in Cassiopeia, showed that there was no detectable parallax, and concluded that it was not 'some kind of comet or a fiery meteor, whether these be generated beneath

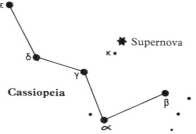

Figure 6: Position of Tycho's Star of 1572.

the Moon or above the Moon, but that it is a star shining in the firmament itself—one that has never previously been seen before our time, in any age since the beginning of the world'. It did not last indefinitely, and as the weeks went by it faded noticeably. By December it was about equal to Jupiter; in March, of the first magnitude; in April and May it sank to the second magnitude, so that it was comparable with the Pole Star. The fading went steadily on, and the last sighting of it was made in March 1574. Its colour changed also, and Tycho was not slow to point out the astrological implications:

'The star was at first like Venus and Jupiter, giving pleasing effects; but as it then became like Mars, there will next come a period of wars, seditions, captivity and death of princes, and destruction of cities, together with dryness and fiery meteors in the air, pestilence, and venomous snakes. Lastly, the star became like Saturn, and there will

Sextant used by Tycho Brahe at his observatory at Hven

A reconstruction of a Copernican/Ptolemaic astrolabe

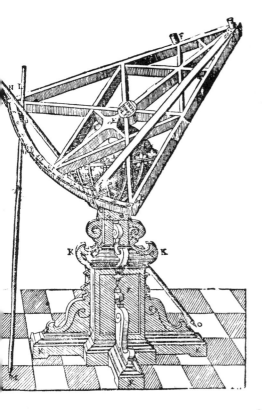

finally come a time of want, death, imprisonment and all sorts of sad things.'

This latter extract, like the first, is taken from the book *De Stella Nova (On the New Star)*, which Tycho wrote and which was published in 1573. Originally Tycho, like Copernicus, was reluctant to burst into print, but his reasons were different. Copernicus feared ridicule and even persecution; so far as Tycho was concerned, the burning question was whether a nobleman of his rank should so far demean himself as to turn into a mere author. Friends persuaded him that it would be quite in order, and the book duly appeared. Yet Tycho had not the slightest inkling of what the star really was. He later suggested that it was formed of 'celestial material', but that this material was less perfect than that of the normal stars, so that it gradually dissolved; he believed that the star faded because of shrinkage, and that by the end of 1573 it

Opposite: *The Milky Way in Cygnus*

The New Star (Supernova) of 1572, centre, from Stella Peregrinæ *by Cornelius Gemma, 1573*

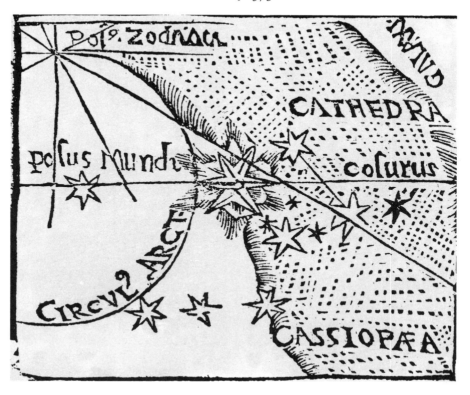

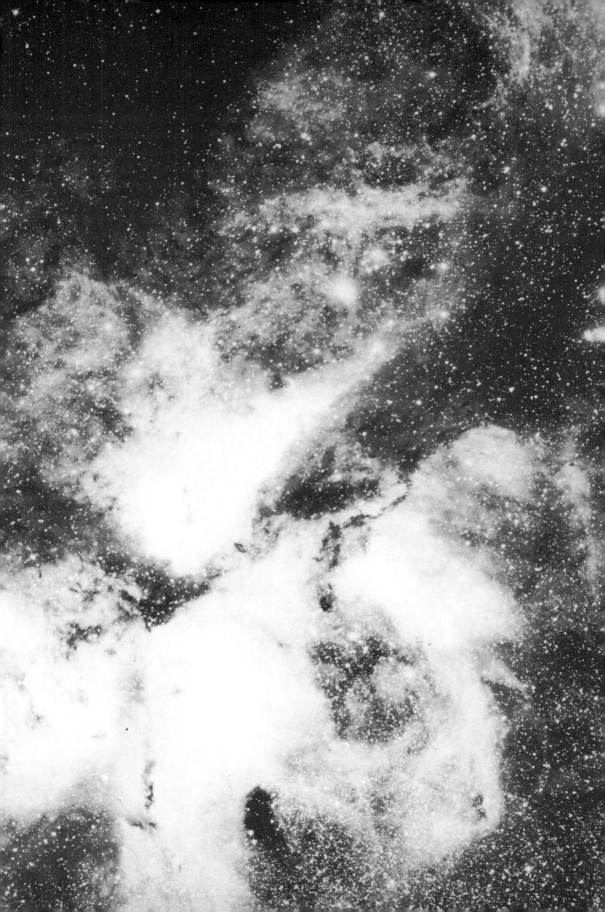

can have been no larger than the Earth. Moreover, he concluded that it was formed out of the Milky Way, and he even thought that after the star had faded away he could see a 'hole' in the Milky Way near the position in which it had burst forth.

If these ideas seem curious, cast your mind back to one Georg Busch of Erfurt, a German who wrote two pamphlets about the star. He regarded it as a comet, 'formed by the ascension from Earth of human sins and wickedness, making a sort of gas which was then set on fire by the anger of God. This poisonous stuff', he added, 'falls down on people's heads, and causes all kinds of unpleasant phenomena, such as diseases, sudden death, bad weather and Frenchmen.' (Comment from the French authorities is not, fortunately, on record.) Other writers held that the star faded because it was travelling away from us in a straight line. These theories were not, in themselves, important; the vital fact was that the star of 1572 made Tycho Brahe decide to devote the rest of his life to astronomy. Had he not done so, the scientific revolution would have taken much longer than it actually did.

By now Tycho was becoming well-known, and it was also about this time that he married—or, at any rate, set up household with the woman who was to remain with him for the rest of his life. Her name was Christine, and she was of humble birth, which did not worry Tycho in the least even though he was so haughty and imperious by nature. We know little about Christine except that from all accounts the arrangement was a happy one, and resulted in the birth of eight children, of whom only two died young. There have been long discussions as to whether the union could be regarded as strictly legal, but at any rate it worked well.

In the next year Tycho gave some lectures in Copenhagen, at the direct request of King Frederik—plus the royal assurance that it was not beneath the dignity of a nobleman to speak in public. Then he undertook some more travelling, and one man whom he visited was Wilhelm IV, the Landgrave of Hesse, who was an enthusiastic and skilled astronomer. Tycho stayed only a week—he left because of the sudden death of one of the Landgrave's daughters, and he did not want to intrude upon his host's grief—and the two never actually met again, but the interlude had far-reaching results. Wilhelm had learned that Tycho had ideas of settling in Germany, and so he wrote to the Danish King: 'Your Majesty must on no account permit Tycho to leave, for Denmark would lose its greatest ornament.' Frederik II saw the point. And on the morning of 11 February 1576, Tycho received a summons to the Court. The King's message was direct and straightforward. If

Tycho would stay in his native land, then he would be given funds and a site for a proper observatory; and Frederik recommended the low-lying island of Hven, which was not far from Copenhagen and is even closer to the modern Swedish town of Malmö.

Hven, known today as Ven, is neither large nor crowded, and it seemed to be ideal as a site for Tycho's proposed observatory. It would have been folly to refuse, and only two days after the King's original offer the grant of the island was completed. Tycho was to be landlord, so that he could collect and use the rents payable by the local inhabitants – something which caused a good deal of trouble in later years. Without delay there began the building of Uraniborg, the 'Castle of the Heavens'. Charles Dancy, the French Minister, laid the foundation stone in August, and within a few years the castle was complete. It was of somewhat Flemish appearance; a square building nearly fifty feet to one

Tycho Brahe's observatory at Uraniborg, Hven, where most of his main work was carried out. After he left Denmark in 1596, the observatory was never used again, and fell into ruins

side, with two semicircular bays attached to the northern and southern façades, and with a maximum height for the main castle of just under forty feet. Quite apart from being an observatory (and the most efficient ever set up in pre-telescopic times) it was also a luxurious nobleman's home, and it even had running water in the bedrooms. Another feature of it was a gaol, which was sometimes used to house tenants who were reluctant to pay their rents. Later Tycho added another observatory nearby; this was called Stjerneborg, and the instruments were placed in underground rooms, so that they were well protected from the wind.

It would take too long to describe the various astronomical instruments. Though none has survived, we know just what they were like, because Tycho left detailed descriptions of them. They were used for making very accurate measurements of the positions of the stars and planets; there were huge quadrants and sextants, all made under Tycho's supervision and some of which were constructed by himself. A workshop was also set up and, slightly later, a printing press.

As the years passed by, Hven became known as a major scientific centre, and there were many learned visitors, some of whom stayed for months or even years. (James VI of Scotland, afterwards James I of England, went there for a day in the spring of 1590, and is said to have presented Tycho with two large dogs.) Another personality was Jep, a dwarf who is said to have sat at Tycho's feet during meals, chattering incessantly – and is said to have been endowed with second sight! Tycho's imperious nature was becoming more and more marked, and he was most certainly a harsh landlord, but so long as King Frederik II ruled the country his position was secure, and he was able to carry out an amazing amount of research.

Of course the main work was on the star catalogue. In all, the positions

Tycho Brahe's notes on the comet of 1577

of 777 stars were measured with real precision; the errors never exceeded four minutes of arc, which is almost incredible when we remember that Tycho had no optical instruments, and had to depend upon his eyes alone when using the giant quadrants and sextants. Also under constant review were the wanderings of the planets; but for the lengthy series of measurements of Mars, Kepler would never have been able to work out his Laws of Planetary Motion. One very interesting piece of research was carried out in 1584, when Tycho checked on the value of the obliquity of the ecliptic as given by Copernicus, and found it to be wrong. This obliquity is the angle between the apparent yearly path of the Sun in the sky (known as the ecliptic) and the celestial equator; it can be measured by finding the noonday altitude of the Sun. Copernicus, thought Tycho, had neglected the effects of refraction, or the bending of light-rays in the Earth's atmosphere, and so he dispatched his assistant Elias Olsen to Frombork to make new measurements. Olsen proved Tycho's point; it was a strange union between the memory of the quiet theorist and the bombastic practical observer. While at Frombork, Olsen was given Copernicus' old triquetrum, which he brought back to Tycho. It had an honoured place in Hven; now, alas, it too has been lost.

During Tycho's stay at Hven, seven comets came into view, the first of which was seen in November 1577. Tycho, of course, observed it and measured it. There was no parallax, as there would have been if the comet had been relatively near at hand. This at once disposed of the old theory that comets were 'atmospheric exhalations' a few miles up, and it hammered the last nail into the coffin of a changeless sky. The vital measurements were made on 23 November, when the comet was close to the bright star Epsilon Pegasi. Tycho made two checks on its position, over an interval of three hours, and calculated that if the comet were as close as the Moon the second distance from the star in the sky would be the same as the first—allowance being made for the comet's own motion. Instead, it was twelve minutes of arc less, and he deduced that the comet must be at least six times as remote as the Moon. Incidentally, it had a retrograde orbit—that is to say it travelled in a direction opposite to that of the Earth. (Many others, including the famous periodical comet named after Halley, do the same thing.)

Other comets showed a similar lack of detectable parallaxes. And then, in 1582–3, Tycho was finally forced to abandon the Ptolemaic system when he found—or thought he found—that the distance of Mars from the Earth was less than that of the Sun, though on Ptolemy's theory it should have been greater. Oddly enough this seems to have stemmed from a misunderstanding on the part of one of Tycho's pupils, who had been carrying out some of the calculations, but it had far-reaching results. The Earth could not be the 'centre of the world' in the way that Ptolemy had thought. But could Copernicus be right in putting the Sun in the position of supreme importance?

Now, Tycho had the greatest admiration for Copernicus; of this there is no doubt whatever. Yet he could not accept Copernicanism. Religious objections were part of the trouble, but there were scientific ones also, and it seems that on the whole the latter loomed larger in Tycho's mind. It seemed to him that the Earth must be too heavy and ponderous to move; also, the lack of star parallaxes meant that if the Earth were in motion round the Sun, the distance between the orbit of Saturn and the nearest stars would be improbably great; and if the apparent diameters of the stars were several minutes of arc, as Tycho believed, a remote star would have to be of incredible size, perhaps as big as the orbit of the Earth. Actually, the stars are immensely more remote than Saturn (represent the orbit of Saturn by a circle six inches in radius, and the nearest star beyond the Sun will be over four miles off) and we do know of stars whose diameters are greater than the modern value of 186 million miles for the diameter of the Earth's orbit. But this could hardly be suspected in the sixteenth century, and Tycho could not realize the true facts. It is quite unfair to claim, as some writers have done, that his objections to Copernicanism were due solely to religious obstinacy.

If both Ptolemy and Copernicus were wrong, then something else had to be worked out, and Tycho came up with what we now call the Tychonic system, even though it was not new. This time the Earth retained its central position, and did not move. Round it travelled first the Moon and then the Sun, with the planets orbiting the Sun at various distances. It sounds rather complicated, and indeed it was, but it was not unreasonable. Neither was there any objection to the crossing of the orbits of Mars and the Sun, since Tycho had no faith in the idea that celestial paths were truly solid spheres (Fig 7).

Unkind people have since called the system 'Tycho's Folly', but the great Dane was inordinately proud of it, and reacted furiously when he believed that it had been copied and re-presented by an avowed enemy of his, Reymers Bär ('Ursus'), who had begun life as a swineherd and had become a professor of mathematics. Tycho was never one to be modest about his achievements. But more mundane matters had begun to intrude upon his life at Hven, and the period of fruitful observation was coming to an end.

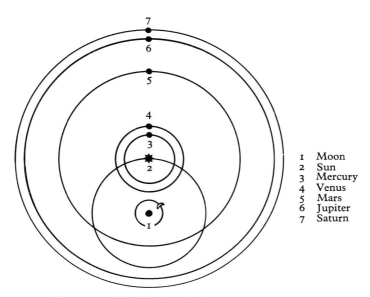

1	Moon
2	Sun
3	Mercury
4	Venus
5	Mars
6	Jupiter
7	Saturn

Figure 7: The Tychonic System.

The trouble began when King Frederik II died in 1588. His son, who became Christian IV, was still a boy (he had once visited Tycho at Hven, and had been warmly greeted) and not all the Danish nobles were well-disposed. Tycho had a friend and ally in the Chancellor, Niels Kaas, but then Kaas died too, and the situation grew more and more awkward. Let us admit, without reservation, that the main fault was Tycho's own. His hot temper, his treatment of his tenants, and his failure to undertake various duties with which he had been entrusted – such as maintaining the famous Kullen Lighthouse – did not help his case in the least. When his main revenue was cut off, Tycho made the crass mistake of leaving Denmark and going to Wandsbeck in Germany, abandoning Hven and taking his retinue and the main observing instruments with him. Another luckless episode was a quarrel with an ex-pupil, Gellius, who was at one time engaged to Tycho's daughter Magdalene. The last recorded observation from Hven was made on 15 March 1596; a great episode was over.

Tycho's next mistake was to write a letter of complaint to the King. Not surprisingly, Christian answered in terse, angry tones; and that was the end of the affair so far as Tycho was concerned. He resigned himself to becoming an exile, and he looked around for a new patron.

He found one in Rudolph II, the Holy Roman Emperor, who was a gloomy, astrologically-minded incompetent. After a spell at Wandsbeck and visits to Dresden and Wittenberg, Tycho arrived at Prague, capital of Bohemia, to take up his duties as Imperial Mathematician. Rudolph made him splendid promises, and installed him at the castle of Benatky, where, it was said, a new observatory would be built. But the situation was very different from that which Tycho had always known at Hven. He was no longer politically powerful, and since the Emperor was chronically short of money there was often no salary either; transport of the instruments was yet another problem, and there are reasons to think that Tycho wished himself back in Denmark. But the die was cast, and he did his best. Before long he left Benatky and came back to Prague. Meanwhile, another important event had occurred: the arrival of Johannes Kepler, a young mathematician from Protestant Germany.

The story of Kepler is best reserved for the next chapter, and for the moment it is enough to say that Tycho was glad to have him as an assistant. There were initial difficulties, because Kepler felt—rightly—that the Dane did not treat him as an equal; but these difficulties were sorted out, and the two men began to plan the compilation of a new set of tables of planetary motion, to be called the *Rudolphine Tables* in honour of their patron. Then, on 24 October 1601, Tycho died after a brief illness. He was still a comparatively young man, but he had not spared himself. Kepler was with him to the end, and Tycho begged him

The Holy Roman Emperor, Rudolph II, who was interested mainly in mysticism and astrology but who encouraged astronomical science

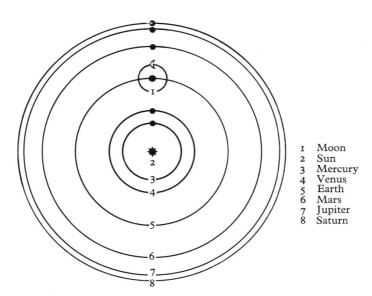

Figure 8: The Copernican System.

1	Moon
2	Sun
3	Mercury
4	Venus
5	Earth
6	Mars
7	Jupiter
8	Saturn

to do two things: complete the *Rudolphine Tables,* and use the mass of Hven observations to show that the Tychonic System, not the Ptolemaic or the Copernican, was the true plan of the universe (Fig 8).

Kepler promised to try. He did indeed finish the Tables, but the Hven material was used for another purpose; it demonstrated that the Solar System was neither Ptolemaic, Tychonic nor properly Copernican.

Unlike Copernicus and Kepler, Tycho Brahe made no single discovery which has made his name immortal. Neither did he, like Galileo, record unsuspected wonders in the sky (it is ironical that he, who would have used the telescope so superbly, was a decade too early for it). But he was the supreme observer, and in this he may never have been surpassed. By modern standards the positions of his 777 stars are of low accuracy; but how many men could have worked them out so well with the clumsy, naked-eye instruments which were all that Tycho had?

There is little more to be said about Hven, which was ceded to Sweden in 1658 and is still Swedish. Uraniborg and Stjerneborg were never used again, and nothing now remains of them. The site of Tycho's castle is occupied by a grassy dip and some trees, with only a few markers to indicate where the building once stood. Yet when I visited Hven a year or two ago, and walked up to the old observatory, I found Tycho's statue—vast, commanding and powerful, surveying the scene where so much pioneer work had been done. The spirit of the Master of Hven still lingers on.

Circles or Ellipses?

Up to now I have been able to tell the story of the scientific revolution in what is, I hope, a straightforward way. Tycho Brahe's life did not overlap that of Copernicus; and though both the next two principal actors—Kepler and Galileo—were growing up during the time when observations at Hven were in full swing, there was little direct contact between Tycho and Kepler until the great Danish astronomer was near the end of his career, while Tycho and Galileo never corresponded at all. For that matter, Kepler never met Galileo face to face, but their stories are so interwoven that they cannot be separated. What I propose to do, therefore, is to discuss Kepler's contribution first, and link Galileo in afterwards. One justification for this procedure is that the main battle between Geocentric and Heliocentric virtually ended with the death of Galileo in 1642, even though murmurings of dissent lingered on until Newton silenced them forever.

It was a strange twist of fate which drew Tycho and Kepler together, even though briefly, because no two men could have been less alike. Tycho was a nobleman, intensely practical, a brilliant observer, and unashamedly egoistic. Kepler came of a somewhat disreputable family background, and had no practical skill at all, quite apart from the fact that a boyhood illness left him with poor sight. He was essentially a theorist, and represents a link between the old and the new, though not in the same way as Galileo. Some of Kepler's views were far in advance of their time, while in other respects he reached back into the mediæval past. It is fair to describe him as a curious mixture.

Johannes Kepler was born on 27 December 1571 at Weil der Stadt in Württemberg. His grandfather had been the local burgomaster, and no

The Belvedere, part of the complex of Prague Castle. Kepler and Tycho made observations from here in 1600–1, under the patronage of Rudolph II

doubt a much-respected local citizen, but the family reputation had declined. Johannes' father was shiftless and irresponsible; he was ill-tempered as well, and he was frequently away from home, as he volunteered for military adventures in the Low Countries and elsewhere. Finally he disappeared altogether, and nobody knows (or cares) what happened to him. Katharine, his wife, was a busybody with a thoroughly malicious tongue and a leaning toward the occult—which was to cause her a tremendous amount of trouble in later life, when she was accused of witchcraft. For Johannes, eldest of the six Kepler children, the home background was far from happy or peaceful, and those early days set the pattern for the life he was to lead. Yet despite illness he soon showed exceptional academic ability, and from this point of view he did well at school, though from all accounts he was never popular with his class-mates.

Only one event of his boyhood merits extra note here. In 1577 his mother took him outdoors to look at the bright comet which had appeared in the sky. This was the comet which Tycho had observed from Hven, finding no parallax and concluding that it must be much farther away than the Moon. Another scientist who made similar

Witches being burnt at the stake from a woodcut circa 1560. Kepler's mother was accused of witchcraft in 1615. She was eventually acquitted after a three year trial

measurements was Michael Mästlin, of the University of Tübingen in Germany, who was to play a major rôle in Kepler's career. It is easy to picture the sinister, witch-like Frau Kepler pointing upward and showing her son this strange apparition glaring down at them, and no doubt this was in Kepler's own mind when, much later, he wrote his fantasy about a voyage to the Moon.

Kepler meant to become a Lutheran pastor, and he went from the local school on to an academy for theologians which was supported by the Duke of Württemberg for the sons of 'poor and pious people'. From here, in 1589, he went on to Tübingen, and his path crossed that of Mästlin, who was to become both his teacher and his lifelong friend. It was during this period that he became a convinced Copernican, inasmuch as he was sure that the Sun, not the Earth, must lie in the centre of the universe. There are reasons for supposing that Mästlin held the same view – but to say so would have been to court dismissal; for official purposes the Ptolemaic theory was still supreme.

Kepler had been aloof and unpopular at school. It was the same at Tübingen, and indeed he never did have the happy gift of making friends easily. Also, he had no money apart from his scholarship grants, since his parents were quite unable to supply any, and so he was cut off from ordinary student life. Then, in 1594, the authorities of the Protestant seminary or boys' school at Graz, in Styria, wrote asking whether Tübingen could supply a promising graduate to teach mathematics and the elements of astronomy. Kepler was nominated, and, perhaps with inward misgivings, he went. It was his first post, and basically it was not a success, because he was a bad teacher; but he continued his own studies, and also dabbled quite energetically in astrology, casting horoscopes and producing calendars which gave assorted information ranging from the dates of eclipses to the best times for collecting medicinal herbs.

It was during his spell at Graz that he began his lifelong quest for what he called the harmony of the universe, and he did his best to find a solution for the rhythm and the design of the Solar System. He did not doubt that the central position was occupied by the Sun, and because the stars showed no parallaxes he had to admit that they were very remote, but he could never bring himself to believe that they were suns in their own right. He considered the Ptolemaic epicycles and deferents, together with eccentric circles – that is to say, orbits which were perfectly circular, but were not centred on either the Earth or the Sun, thereby allowing for variations in the observed distances of the planets.

He also thought hard about another idea. Once it had been established that a planet moves in a way which makes its distance from the Earth vary, it had been suggested that each planet must have a 'sphere' or layer of space in which it must always be found. So that no space should be wasted, it was added that these spheres must touch each other; thus the outer boundary of Jupiter's sphere (i.e. the greatest distance of Jupiter from the Earth) must touch the inner boundary of the sphere of the next planet, Saturn. Unfortunately for the neatness of the theory, Copernicus had shown that this is not true, and Kepler tried hard to solve the problem. He considered the five regular solids of geometry, from the cube (with six equal faces) to the icosahedron (bounded by twenty equilateral triangles), and fitted them in with the spaces between the orbits of the six planets – Mercury, Venus, the Earth, Mars, Jupiter and Saturn. He found what he believed to be a valid relationship, and developed a complicated pattern which can be better drawn than described. It sounds – and is – much less logical than anything which Ptolemy had produced, but Kepler was convinced by it, and always regarded it as the greatest of his discoveries.

Prague, from an engraving by Filip van der Bosache, 1606

He announced it in a book, published late in 1596 and seen through the press by Michael Mästlin. We remember it as the *Mysterium Cosmographicum* or *Cosmographic Mystery*, though the full title was much longer. *(Prodromus Dissertationum Cosmographicarum de admirabili Proportione Orbium Cælestium deque Causis Cælorum numeri, magnitudinis motuumque periodicorum genuinis et propriis, demonstrarum per quinque regularia corpora Geometric.)* Let us admit, in retrospect, that the book was of scant scientific value, and is no more than an historical curiosity, but it had a profound effect on Kepler's career, because he sent a copy to Tycho Brahe and received an encouraging reply. Not that Tycho had much faith in the significance of the five regular solids, but he did realize that this unknown young German was a mathematician of no mean skill.

Galileo, in Padua, also received a copy of the book, and replied politely, saying that he was looking forward to reading it. Whether he ever did so is by no means certain. Yet he did say, in his letter to Kepler, that he had long been a believer in the Copernican theory, but had not said so openly 'alarmed at the fate of our master Copernicus himself who, although he has attained immortal renown in the eyes of some, yet, in the eyes of a countless host (so great is the number of fools) he appeared

as one to be laughed at and hissed off the stage'. Kepler wrote back, urging him to make his opinions public. But Galileo was not ready to do so, and the two great men had no further contact for many years afterwards.

There were two other ideas in the *Mysterium* which should be referred to before we go any further. One was in connection with the movements of the planets round the Sun. According to the teaching of the time, movement will eventually stop unless a constant force is applied, and so Kepler had to suppose that the Sun exerted some regular 'pushing' force which whirled the planets round; it could not be visible light, but was something whose nature remained unknown, and whose effects fell away with increasing distance, thereby explaining why the most remote planet, Saturn, was the slowest-moving member of the system. Also, Kepler believed that the universe began with the planets stretched out in a line from the Sun out to the origin of the Zodiac, perhaps four thousand years or so earlier. It was, he thought, possible that the world would end when the next similar lining-up took place, though he had no idea of when this would be.

Kepler married in 1597, and although the union was by no means blissfully happy it did not break up. Unfortunately, religious troubles were looming ahead. Under the régime of the Archduke Ferdinand, the Protestants in Graz were put under all sorts of pressure, and suddenly, in September 1598, all Lutheran teachers were ordered to leave the city at a day's notice, under pain of death. Kepler went, no doubt wisely. Subsequently he was allowed back, but the situation was obviously most uncertain, and he looked around for a safer home. There was no post for him at Tübingen, despite the continued friendship of Michael Mästlin, and so he accepted the invitation to join Tycho Brahe at Prague. He made one final visit to Graz, hoping that he would be able to retain some kind of a foothold there even though living in Prague; but Ferdinand dismissed him together with all other leading Protestants, and told him firmly that he had no more than six weeks' grace. So Kepler fled back to Prague, depressed and with no assured prospects. It would be too much to hope that two men of genius and yet with such different temperaments could work together for long; Tycho was never easy in his personal relationships.

In fact the matter was never put to the test, because Tycho died soon afterwards and Kepler succeeded him as Imperial Mathematician—at a much lower salary, let it be said. Things were not easy. Even the meagre amounts of money which he had been promised were never paid regularly, and Kepler had to do all the routine work himself instead of having the help of assistants. Also, Tycho's son-in-law, Tengnagel, claimed all the instruments and manuscripts, and there were constant wrangles which went on for the rest of Kepler's life. The Emperor Rudolph II was peculiarly unfitted for the rôle in which Fate had cast him, and it was only a question of time before his blundering incompetence led to his removal from power, though in fact he held on to his authority until 1611. So Kepler had to work under difficulties, and it is truly remarkable that he managed to accomplish as much as he did.

He had, however, one priceless asset: the legacy of Tycho's observations made at Hven. He had been charged with using them to prepare new tables of planetary motion, and to demonstrate the truth of the Tychonic system of the universe. Obviously his first task was to work out a proper 'world system', and, try as he might, none of the accepted principles seemed to work – unless he were prepared to believe that some of Tycho's observations were faulty. Kepler was not prepared to make any such concession. He knew that the great Danish astronomer would never be guilty of errors so large as this, and he continued his attempts to make sense out of the observations which he had. The vital work was connected with the movements of the planet Mars, to which Tycho had always paid special attention. We know now that this was a lucky choice, for reasons which will soon become evident.

Nebula NGC 6960 in Cygnus

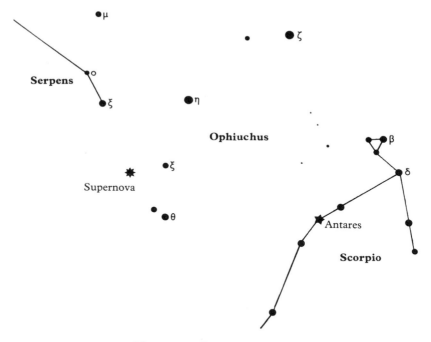

Figure 9*a*: Position of Kepler's Star.

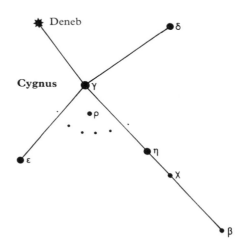

Figure 9*b*: Position of P Cygni.

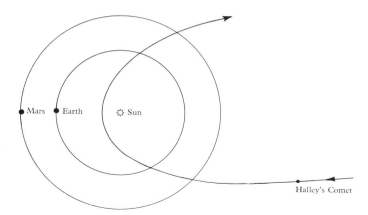

Figure 10: Orbits of the Earth, Mars and Halley's Comet.

During this trying period another bright new star blazed out, this time in the constellation of Ophiuchus, the Serpent-Bearer (formerly called Serpentarius). It was first seen on the night of October 10–11 1604 by a Court official named Johann Brunowski, who was a keen amateur astronomer and who told Kepler about it. Kepler was sceptical, but as soon as the clouds cleared away from the Prague sky, on 17 October, he saw it for himself; by then it had already been found by astronomers elsewhere, including Mästlin, still lecturing at the University of Tübingen. It was as bright as Jupiter, and it faded only slowly; Kepler last recorded it in 1606.

Kepler, of course, did not believe the stars to be suns, and he followed Tycho's ideas about the star of 1572; the newcomer was made up of matter from the Milky Way, which gradually dispersed. In his book *De Stella Nova* (the same title as Tycho had used!), published in 1606, he threw in some astrology; the star indicated the approaching destruction of the Turkish Empire, and so on. There is also an interesting note about yet another new star, seen in the constellation of the Swan in 1600 and which became clearly visible to the naked eye without ever growing really conspicuous. We know now that the Swan star – P Cygni – was not an ordinary nova at all, but merely an irregularly variable star of unusual type; it is still visible with the naked eye, and the diagram (Fig 9) shows its position near the central star of the cross of Cygnus. But the brilliant newcomer of 1604 was another supernova, strictly comparable with Tycho's. No supernovæ have been seen in our Galaxy since then, but we can still pick up long-wavelength radio waves from the débris of Kepler's Star. (There has been one supernova visible with the naked eye; this flared up in 1987 in the external system we call the Largo Cloud of Magellan. It became quite prominent, even though it was 169,000 light-years away, and remained a naked-eye object for weeks before fading back to obscurity.)

At least it was shown that the 1604 supernova showed no parallax, though it is unlikely that Kepler made any reliable measurements himself. His book caused a great deal of interest, but many of the comments and criticisms about it were astrological, and there is no need to discuss them here.

Meanwhile, Kepler was still trying to find a way to make Tycho's observations of Mars agree with some kind of logical system. There was one extra problem. On the old geocentric theories, the Earth was in the centre of the universe, but not at the actual centre of the orbits of the planets, which was some distance away. On the newer heliocentric theory, it had to be assumed that the centre of the planetary system was some way away from the Sun. But this would not fit in with Kepler's own concept that the planets were being whirled around by some unknown force which was truly centred in the Sun. No kind of circular motion would do; combinations of circular motions were equally unsuccessful. At last Kepler stumbled upon the truth. The planetary orbits were not circular at all; they were ellipses.

I say that Kepler 'stumbled upon the truth' because, as he himself realized later, he had had the solution in his grasp for some time. Moreover, the orbit of Mars is more eccentric than those of most of the other planets, and it was Mars upon which Kepler had been concentrating. Altogether there was a considerable element of luck in his final triumph though this does not in the least detract from his skill, his patience, or his commendable (and justified) faith in the accuracy of the observations left by Tycho Brahe.

Before going any further it will, I think, be helpful to say a little more about ellipses in general, because many people have the wrong idea about the shape of the Earth's path. Our world moves round the Sun in an orbit which is almost circular, but not quite. The distance-range is between $91\frac{1}{2}$ million miles (in December) to $94\frac{1}{2}$ million miles (in June); the seasons, incidentally, have very little to do with this changing distance, and are caused by the tilt of the Earth's axis, which Tycho had measured so carefully. The northern hemisphere is tilted sunward in June, so that Europe and the United States receive the full benefit of the rays, while in northern winter it is the turn of Australia to be tipped toward the Sun. Draw the Earth's orbit to scale, as shown (Fig 10), and it appears to all intents and purposes circular. The next figure shows the orbit of Mars, also to scale; the range here is between $128\frac{1}{2}$ million miles and $154\frac{1}{2}$ million miles, so that the departure from circularity is more noticeable. And in the third diagram we have the orbit of Halley's Comet, where the distance ranges between 56 million miles at closest approach (perihelion) out to almost 3300 million miles (aphelion). In each case the Sun occupies one focus of the ellipse, while the other focus is empty. The revolution periods are 76 years for Halley's Comet, 687 days for Mars and, of course, $365\frac{1}{4}$ days for the Earth.

I give these diagrams mainly to show that the Earth's path is not markedly oval, as some people fondly imagine. It is so nearly circular that the difference may seem unimportant. In fact, it is of the utmost significance, and one is lost in admiration both for Kepler, who worked it out 'from scratch', and for Tycho, whose essential observations had to be made without any optical help at all.

The idea of elliptical orbits was not entirely new – it had been proposed, rather tentatively, by Arzachel of Toledo as early as 1080 – but it was revolutionary in every sense of the term, and even some of Kepler's admirers found it hard to accept. One of these was David Fabricius, a

Engraving of the constellation of Serpens from Kepler's De Stella Nova, *published in 1606*

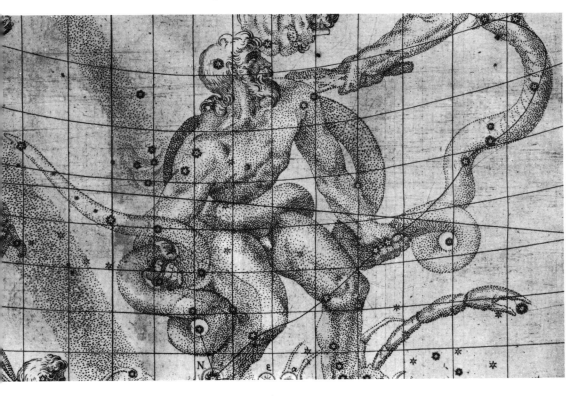

119

skilled observer whose measurements of the position of Mars for the years 1602 and 1604 had actually been used to supplement Tycho's series. Galileo, as we will see, never did believe in it. But the main breakthrough had been made, and Kepler was able to draw up the first two of his classic Laws of Planetary Motion, publishing them in a book, *Astronomia Nova (New Astronomy)* which was actually finished in 1606 although not published until 1609. The third Law was added in 1618, but for the sake of clarity it is best to give them all together. They are as follows:

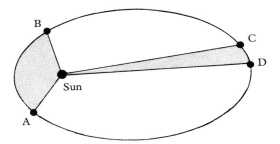

Figure 11: Kepler's Second Law.

1. A planet moves in an elliptical orbit, the Sun occupying one of the foci.
2. The radius vector – that is to say, the imaginary line joining the centre of the planet to the centre of the Sun – sweeps out equal areas in equal times.
3. The cubes of the mean distances of the planets from the Sun are proportional to the squares of their periods of revolution.

Law 1 needs no further clarification. Law 2 is almost equally straightforward. It means, in effect, that a planet (or a comet, for that matter) will move at its quickest when it is nearest to the Sun. In the diagram, (Fig 11) our planet moves from A to B in the same time that it takes to move from C to D, S being the Sun. Then the sector ASB will be equal in area to the sector CSD. We know that this is because of the effects of gravitation; Kepler attributed it to the falling-away of his 'whirling force' with increased distance from the Sun, but the result was the same.

Now let us turn to Law 3, which, I am glad to say, can be explained by using no more than the very simplest mathematics. Consider the Earth and Mars. Taking the Earth's mean distance from the Sun as 1 unit, that of Mars is 1.523. (The real mean distances are 93 million miles and 141.5 million miles respectively; and if you multiply 93 by

1.523, you will find that the answer is 141.5.) The periods are in the ratios of 1 to 1.88. (The real periods are 365.25 and 687 days respectively; in round figures, 365.25 multiplied by 1.88 comes to 687.) The cubes of 1 and 1.523 are 1 and 3.54, and the squares of 1 and 1.88 are also 1 and 3.54.

It follows that as soon as you know a planet's period of revolution round the Sun, Kepler's third Law gives an easy way of calculating its relative distance from the Sun, so that we can build up a complete scale model. This, in fact, is how the size of the Solar System was worked out. But you have to know one *absolute* distance; and Kepler did not. He believed that the Sun could be no more than 14 million miles from us. This was an improvement on the estimates of Tycho (5 million miles) and Copernicus (2 million), but it was still far too low, and the first reasonably good measurement was not made until 1672 – by the Italian astronomer Cassini, who gave 86 million miles.

But again I am running ahead of my story, so let us return to Kepler's *Astronomia nova*. It did not arouse nearly so much excitement as might have been expected, perhaps because it was printed in a limited edition and was not easy reading, but it contained all manner of new ideas as well as the first two Laws. There are even some glimpses of modern views on gravitation. 'If two stones were placed in any given part of the universe . . . then the two stones, like two magnetic bodies, would unite at some intermediate point, each approaching the other through a distance proportional to the mass of the other. . . . The sphere of the Moon's attraction extends to the Earth and draws up water in the torrid zone.' Here we have some remarkable insight, and a preview of Newton's discoveries about the tides. Kepler knew that the more massive a body, the stronger its effects, and he added: 'If the Moon's attractive force extends as far as the Earth, then much more must the Earth's attractive force extend as far as the Moon, and beyond; moreover, nothing consisting in any way of earthly material, caught up on high, can escape from the powerful hold of this attraction.'

There was one more interesting comment about the mysterious force which, Kepler believed, came from the Sun and propelled the Earth and the other planets along. It could not, he said, be identical with light – or else the Earth would stop in its tracks whenever the sunlight happened to be cut off during a solar eclipse!

Now, for the first time, it was possible to produce really accurate tables of the movements of the planets, though much work remained to be done. One episode shows both Kepler's skill as a mathematician and his inadequacy as an observer. He predicted that on 29 May 1607

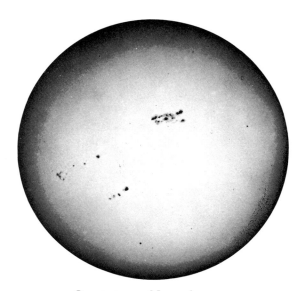

Sunspots, 15 November 1970

the planet Mercury would pass directly between the Earth and the Sun, thus appearing in transit as a dark spot against the brilliant solar disk. He made a screen with a small hole, and passed the Sun's light into a darkened room; there, on the solar image, he saw a speck—and was convinced that he had seen the transit. In fact, what he had seen was one of the dark, cooler patches which we call sunspots; and if he had checked a few hours later he would have seen his mistake, as the spot would still have been there. Unfortunately he did nothing of the kind, and so missed the chance of discovering sunspots himself. When the spots had been described, by Galileo and others some years later, Kepler realized what had happened. (Let me add, though, that he predicted another transit of Mercury for 1631; and though he was dead by then, the transit was duly observed by the French astronomer Pierre Gassendi.)

Comets remained a problem. Kepler observed one in 1607—we now know it to have been the famous Halley's Comet—and several others later, including three in 1618. He even wrote a book about them, defending the idea that because of their lack of parallaxes they must be very remote; here he disagreed with Galileo, who still believed them to be contained in the Earth's upper air. But Kepler was bothered by the apparent ability of comets to pass through the spheres of the planetary orbits, and he concluded that they must travel in straight lines. They were, he added, lit up by the Sun. Originally he thought that the planets must be self-luminous, or else Venus at least would show phases like those of the Moon, but when Galileo's telescope showed phases of exactly this kind Kepler admitted that he had been wrong.

Of course this leads us on to the invention of the telescope, and its applications to astronomy, which altered the whole situation beyond recognition. But since this belongs more properly to the story of Galileo I propose to defer it for the moment, and will add only that Kepler was most enthusiastic about these new developments; he even designed a new kind of optical system which proved to be a vast improvement on Galileo's.

Meantime, things in Prague had been getting worse and worse. The Emperor Rudolph lost his grip completely, and was compelled to abdicate in favour of the Archduke Mathias. In the same year, 1611, Kepler's son Frederick died, and then his wife Barbara died too. Kepler had no wish to stay in Bohemia, and he managed to obtain a post as District Mathematician at Linz, in Austria. Though he found few congenial companions there, he stayed for fourteen years, during which time he married again; Mathias, now Emperor, actually came to the

Kepler's illustration to explain his discovery of the elliptical orbit of Mars from his Astronomia Nova, *1609*

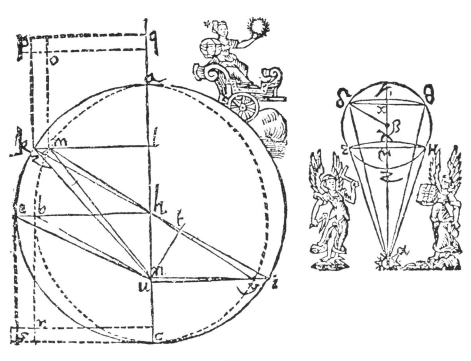

wedding. I will say nothing here about his various minor activities, such as his authorship of a treatise about making accurate measurements of the amount of wine in a cask(!), but I must mention the trial of his mother Katharine, because it took up so much of his time and caused him so much anxiety.

Witches were much feared in the Europe of the early seventeenth century. If identified, they were subjected to the roughest of treatment. Recall the lines written so long afterwards by W. S. Gilbert, in *Ruddigore*:

> Sir Rupert Murgatroyd, his leisure and his riches
> He ruthlessly employed in persecuting witches.
> With fear he'd make them quake; he'd duck them in his lake —
> He'd break their bones with sticks and stones,
> And burn them at the stake.

The fictional Sir Rupert had many predecessors who were only too real, and Kepler's mother laid herself open to accusation. In her old age she was even more interfering and disagreeable than ever, and things came to a head in 1615, when she was brought up before a court in her home town and threatened with death unless she incriminated herself by casting a spell strong enough to cure a woman whom she had allegedly bewitched. After various developments she fled to Linz to join her son, out of the power of that particular court, but then made the mistake of returning home, where she was promptly arrested again. Kepler at once went to her help. The trial lasted for three years, and at one time the old woman was shown the instruments of torture, though she was not actually harmed. It was only her stubbornness which saved her; she refused to admit anything, and eventually she was acquitted, though she died in the following year. All this added to Kepler's troubles, which were already considerable enough. The Emperor Mathias had died, and was succeeded by no less a person than Ferdinand, who had expelled Kepler from Graz long before. However, he took no immediate action now, and it was while Kepler was still at Linz that he published two more books: his *Harmonice mundi (Harmony of the world)* and *Epitome Astronomiæ Copernicæ (Epitome of Copernican Astronomy)*.

The first of these, finished in 1619, was the usual medley of brilliant science and mystical speculation. It contained the all-important Third Law; it also contained more details about the music of the spheres, the five regular solids, and similar topics. The second, the *Epitome*, was really in the nature of a textbook, and summed up the views which Kepler held in this latter part of his life. The Sun is unique, and occupies the centre of the whole universe; round it moves the family of planets,

Opposite: *The Church of the Virgin, Prague, where Tycho Brahe is buried*

and much farther away lies the sphere of the stars. The stars are at different distances from us, as is indicated by their range in brightness, but they cannot be infinite in number, and so they can occupy only a finite part of space. Nothing can lie beyond, for one cannot have space where there are no bodies; the outer boundary of space may also be a solid crystal sphere. This also shows that the daily rotation of the sky is due to the real movement of the Earth, not the heavens; because if the universe is finite, it cannot move (there is no standard of reference) and if it is infinite, the outer parts would have to spin round us at an infinite speed (which is absurd). The variation in a planet's velocity as it is whirled around the Sun by the unknown 'pushing force' can be explained if the core of each planet is divided into two hemispheres, one of which is attracted by this force and the other repelled; these hemispheres maintain a fixed direction in space, so that each hemisphere is turned alternately toward and away from the Sun, causing the planet to be alternately pulled in and thrust outward.

Kepler was never a 'popular' writer in the sense that Galileo had become, but at least the *Epitome* was intelligible to the layman. Once it had been completed, Kepler turned his attention to the one great task which remained: the completion of the *Rudolphine Tables*, which he and Tycho had planned together. But even when they had been drawn up, there were still difficulties about publication. Tycho's heirs continued to be troublesome, and moreover the Emperor was anxious for the tables to be printed in Austria. Kepler did his best, but the printing facilities at Linz were primitive, and everything was painfully slow.

By 1625 the Protestants were again under pressure. As before, Kepler was given special protection, but the outlook was far from reassuring, and when the printing press was destroyed in a riot he at last obtained permission to leave. In the following year he made his final exit, together with his wife, his family and what possessions he had managed to save. He came to the city of Ulm, but found little satisfaction there either, and at one stage he even set out to walk to Tübingen—where some of his happiest times had been spent—though the snow and the cold drove him back. It must have been a profound relief to him when the *Rudolphine Tables* finally came out. After some last bickering with Tycho's heirs, they became available to the world in early 1628.

Opposite top: *The Golden Street, near Prague Castle, where mathematicians,
astronomers and alchemists lived, at the time of Tycho and Kepler*

Opposite bottom: *The Strahov Library, almost unchanged since the time of
Kepler and Tycho, and where some of their printed books and manuscripts are kept*

His next and, as it proved, last migration was to the Duchy of Sagan
in Silesia, whither he went at the invitation of the famous soldier-
statesman Wallenstein, whose horoscope he had cast long before.
Wallenstein was a firm believer in astrology, and he was glad to have
Kepler's counsel. At Sagan, Kepler duly carried out the duties assigned
to him, and also prepared yearly almanacs of celestial phenomena and
meteorological forecasts—in which he was greatly helped by a mathematician named Jacob Bartsch, who subsequently married his daughter

*Kepler's explanation of the structure of the planetary system using the five
regular solids between the spheres of the various planets, from his* Harmonices
Mundi, *1619*

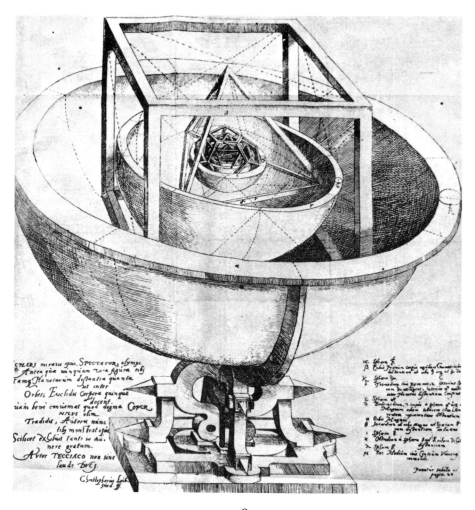

Susanna. But in 1630, with war raging apparently everywhere, Wallenstein fell from power, and the future looked bleak. Leaving his family at Sagan, Kepler set out on a journey to collect some of the money which he was owed, but when he reached the town of Regensburg he collapsed with fever and exhaustion. Medical treatment was ineffective, and on 15 November 1630 he died. It was a sad end to a sad life.

No account of Kepler would be complete without a reference to his famous *Somnium* or *Dream*, which is ostensibly science fiction, but is in effect a spirited defence of the Copernican theory. The first draft was written around 1608, and Kepler returned to it twenty years later, but it was not published until 1634, when his son Ludwig saw it through the press. The footnotes, purely scientific, are much longer than the story, and are highly illuminating.

The hero of the *Somnium* is a young Icelander, Duracotus, whose mother Fiolxhilda was a 'wise woman' who earned her living by

Johannes Kepler 1571–1630, frontispiece from : Rudolphine Tables, *Ulm, 1627*

providing sailors with bags of magical herbs (one need not be very far-sighted to realize the connection with Kepler's witch-like mother, who had showed him the comet of 1577). One day Duracotus tampered with one of the bags, and his mother was angry enough to hand him over to a sailor, who sailed across to Denmark and dispatched the young man to take some letters to Tycho Brahe. After spending five years on Hven, Duracotus returned to Iceland and was reconciled to his mother, who at last made up her mind to show him some of her secrets. She revealed that her teacher was a demon, who lived in the world which we call the Moon but which she knew as Levania.

Both Levania and Volva (the Earth) contain demons, as Fiolxhilda explained, but normally they cannot cross from the one world to the other, because they hate light, and the rays of the Sun are too brilliant. However, during an eclipse the Earth's shadow stretches across the gap, making a 'bridge of darkness', so that for an hour or two the demons can cross at will. Duracotus wanted to make the journey for himself, so Fiolxhilda summoned the demons of Levania. Then, said Duracotus, 'withdrawing from me into the nearest cross-roads, then returned; and commanding silence with the palm of her right hand outstretched, sat down near me. Scarcely had we covered our heads with a cloth, as is the custom, when behold, there came the sound of a voice . . .'

Thereafter the tale becomes a mixture of fantasy and seventeenth-century science. To ease the discomfort of travelling above the air, Duracotus was given an anæsthetic or 'dozing draught', and also sponges, which were moistened and held to his nostrils. (Short of equipping his hero with a space-suit, Kepler could hardly have done much better.) The demons' part was done when they had pulled Duracotus up to the point where the Moon's gravity balances that of the Earth. After this they simply let go, and allowed their passenger to fall moonward on his own. We have here the first indication of the alleged 'neutral point' between the two bodies, later to be used—or, rather, misused—by no less a writer than Jules Verne. Moreover, Duracotus relates that as soon as he reached the neutral point his limbs curled up like those of a spider. 'When the attractions of the Moon and the Earth equalize each other, it is as though neither of them exerted any attraction. Then the body itself, being the whole, attracts its minor parts, its limbs, because the body itself is the whole.' Kepler is describing gravity, even though without any clear understanding of it.

It was already known that the Moon always keeps the same face turned toward the Earth, because its revolution period is the same as its

Frontispiece of the Rudolphine Tables. *These tables represented Kepler's last astronomical work and were so named in honour of Kepler's old benefactor, Rudolph II. (On the pedestal, front section, a map of the Island of Hven, where Tycho built his observatories)*

axial rotation period (27.3 Earth-days in each case). Duracotus tells how Levania is divided into two zones, Subvolva and Privolva. From Subvolva, the Volva or Earth is always visible; from Privolva, never. Levania is a world of extremes, with violent changes of climate, towering mountains and deep valleys. 'The hollows of the Moon first seen by Galileo are portions below the general level, like our oceans, but their appearance makes me judge that they are swamps for the greater part.' The Moon-folk are not human; some are serpentlike, while others have fins to help them swim, and others crawl along the ground. Most are covered with fur. When a Moon creature is unwise enough to allow itself to be caught in the open air near midday its outer fur is singed by the intense heat, so that the creature drops as though dead; at nightfall it revives, and the burned portions of its fur simply fall away. Government cannot exist in the terrestrial form, because the Moon-dwellers have such brief lives. In Subvolva, particularly, the creatures and plants are of monstrous size, but live only for a few lunar days. . . . In the end Kepler cheats his readers – he tells how he awoke to find 'his head covered with a cushion, and his body tangled in a rug'. This is why the book is called the *Dream*.

Quite clearly Kepler was keeping to the scientific facts as he knew them – apart from the demons, needless to say. There was no reason to doubt that the Moon might be inhabited, and the very lengthy footnotes to the text make it clear that what he is doing is to write nothing more nor less than a layman's account of the heliocentric universe.

This, in its way, is typical of him. He, more than any other pioneer, had one foot in the ancient world and the other in the modern. Had he been born of a rich and noble family, his thoughts might have been directed along different lines – but then, he would not have been Johannes Kepler.

Galileo, the Experimenter

We now come to the fourth in our list of geniuses who made the scientific revolution possible. Galileo Galilei was not only one of the greatest of all pioneers of science, but he was also one of the most colourful—even if he did not have a Tycho-like artificial nose. His career overlapped that of Tycho; he was born in 1564, and during Tycho's Hven period Galileo was busy lecturing first at the University of Pisa and then at Padua, but apparently the two never corresponded at all. I have often wondered what would have happened if they had met face to face. They might have respected each other enough to become the closest of friends, though more probably there would have been a verbal explosion of supernova violence. In any case, such a confrontation would have been absolutely fascinating.

But Tycho lived in Denmark and then in Bohemia; Galileo spent his life in Italy. At one stage, moreover, Galileo went to some pains to discredit Tycho's theories in a rather unnecessarily forceful manner. He did have some correspondence with Kepler, but it was very limited, and the fact that it lapsed was not Kepler's fault. It must be said, too, that Galileo, was not always ready to give credit and praise to his colleagues when he ought to have done so, and his attacks on his opponents were stinging as well as witty. One has the impression that he took great pleasure in 'scoring a point', though he did not in the least appreciate having the tables turned upon him (which, from a scientific viewpoint, was admittedly seldom).

Today, Galileo is best remembered for two things: his pioneer use of the telescope in astronomy, and his spirited championship of the heliocentric or Copernican system, which ended in his trial and condemnation by the Inquisition in Rome. There is no doubt at all that he embarked upon a campaign of pro-Copernican propaganda, and that he made some serious miscalculations in the process; tact was never his

133

strong point, and all through his career he made plenty of enemies. He made no single fundamental discovery, and in some ways his ideas were much less advanced than those of Kepler; in particular he could never break away from the concept of perfectly circular orbits, even after the publication of Kepler's Laws, and one of his so-called 'proofs' of the motion of the Earth was based upon a theory of the tides which could hardly have been more wrong. Yet unlike Kepler, he used methods of investigation which had no roots in the ancient world, and he was a born experimenter. Quite apart from his work in astronomy, he made striking contributions to other branches of science, and he has been called the true founder of experimental mechanics.

When I say that Galileo made no single fundamental discovery in astronomy, I do not for one moment mean to belittle his achievements. Though he did not invent the telescope, and was not the first to turn it to the sky, his patient, skilful programme of observation put him in a class of his own. And despite his mistakes, it was Galileo, more than any other man, who showed the world that the Sun rather than the

The University of Padua, at the time of Galileo, circa 1595

Earth is the centre of the planetary system. The fact that he was officially silenced by the Church, first in 1616 and then after the notorious trial of 1633, is neither here nor there.

It is also worth considering why Galileo was subjected to persecution in a way which his famous colleagues were not. Part of the reason is that he was an Italian, and the Italy of those times was no place for anyone with unorthodox views about religion. Galileo himself was a good Catholic, but he had the vision to see that the moment had come

Monument to Giordano Bruno, Rome

to separate religion from pure science, which made him vulnerable at once. Copernicus had come to the same conclusion, but was prudent enough to withhold publication of his great book until he was out of the reach of any potential enemies; Tycho, of course, was in no danger at all, because his system of the universe retained the central Earth and did not conflict with orthodox teaching; Kepler lived beyond the immediate jurisdiction of Rome, and in any case most of his books were addressed to his fellow scientists rather than to the general public. But Galileo's *Dialogue,* the main cause of his falling foul of the Inquisition, was written in Italian rather than Latin and was intelligible to the non-specialist, quite apart from the fact that it was taken to be a personal insult to the Pope. Therefore, said the Holy Office, it was dangerous and had to be suppressed.

In the end Galileo was neither tortured nor put to death. An earlier revolutionary thinker, Giordano Bruno, had been less fortunate. Bruno, fiery of temperament and outspoken in speech, believed the stars to be suns lying at a tremendous distance from us; he was convinced of the truth of the Copernican system, and he discussed it openly, not only in France and England (which did not matter) but also in Italy (which did). He even commented that Osiander's preface to the *De Revolutionibus* was written by one ignorant ass for the benefit of other ignorant asses. He made the tactical blunder of returning to his native country after a lengthy exile abroad, and was, predictably, arrested. Following a mock trial by the Inquisition, he was burned at the stake in Rome in 1600. It would be wrong to suppose that this was due entirely to his defence of Copernicus; in the eyes of the Church he had committed many other crimes as well, but his fate was highly significant, and one may doubt whether Galileo took sufficient heed of it. At the time, he was living in Padua—not so very far away from the Eternal City—so that presumably he knew what had been happening.

Galileo was born on 15 February 1564, in Pisa. His family hailed from Florence, and his father, Vincenzio Galilei, was well-educated and talented; he was, in particular, a very capable musician. He became a trader for purely financial reasons, and this was why he moved from Florence to Pisa some years before the birth of his eldest son. Galileo had two younger brothers and four sisters. One of the brothers, the shiftless and irresponsible Michelangelo, was to cause Galileo considerable anxiety in later life.

The family stayed in Pisa until Galileo was ten, and then went back to Florence, so that it was in these two cities that the future scientist had his first schooling. Presumably he did well, and in the autumn of 1581 he went to the University of Pisa as a medical student. This was his father's wish, but he never pretended to have any real interest in doctoring; his leaning was toward mathematics and mechanics, and unfortunately there was no serious department of mathematics at the University. During a vacation period at home, two years later, Galileo went secretly to a family friend, Ostilio Ricci, for mathematical lessons. When Vincenzio discovered what was going on, he wisely changed his mind. If the boy wanted to be a mathematician—well, he could have his wish.

There was one incident which probably had a profound influence on Galileo's career, and this episode, unlike so many of its kind, seems to be true. During a service in Pisa Cathedral, he watched a lamp swinging from the ceiling at the end of a long cord. He noted that no matter

A portrait of Vincenzio Viviani

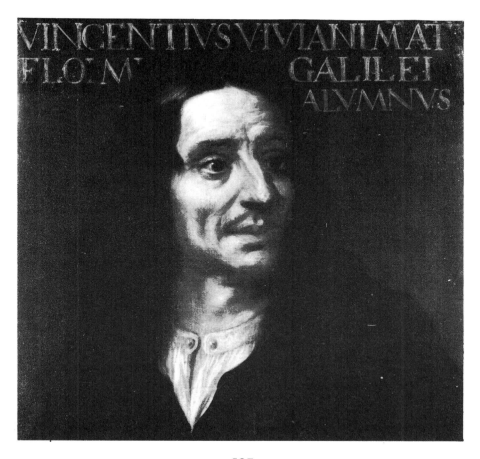

whether the swing were short or long, the time needed for a full up-and-down motion remained the same; in other words, the time of swing is independent of the amplitude, and depends only upon the length of the cord. Needless to say, this could have been noticed by anybody, but Galileo began to wonder just what it meant. Typically, he carried out some practical experiments in order to prove his point, and then he put it to use, working out a device to measure short time-intervals such as a man's pulse-rate. This, at least, is what we are told by Galileo's first biographer, Viviani, though there is no independent confirmation. Whether or not he built such a device, there is the certainty that he learned a great deal from the behaviour of the swinging lamp – more than he would have done by listening to the service!

Galileo's first spell at Pisa ended rather abruptly in 1585. He had no degree, but his father could no longer afford to pay the University fees, and so he went back to Florence, where he spent the next four years. Lacking a definite profession or qualification, a lesser man might have been somewhat daunted, but not Galileo. He had plenty of contact with the cultural groups in which his father was involved, and he continued with his studies of mathematics and mechanics. What he learned, and what he found out for himself, made him more and more suspicious of the old ideas, notably those of the great Greek philosopher Aristotle, who was still regarded as the supreme authority even though he had died seventeen centuries before. Galileo had much more respect for Archimedes, who had been a practical experimenter as well as a theorist (who does not know the story of how he put King Hiero's crown into a bathful of water, measured its specific gravity, and proved that its maker had put 'base metal' in with the gold?). Anyone who questioned Aristotle was regarded with distaste, but as yet there was no hint of trouble. Nobody was likely to bother about the unformed ideas of a scientific dabbler in his early twenties.

However, Galileo began to make his mark. In 1586 he invented a device known as a hydrostatic balance, used for measuring the specific gravities of bodies (shades of Archimedes!) and wrote a short book about it, though the book was merely circulated among his friends and acquaintances instead of being officially printed. He also produced some technical studies about the centres of gravity of solids, and this drew him into correspondence with some prominent scientists, notably a German Jesuit named Christopher Clavius. In rather different vein, it is said that he gave two public lectures about the shape, size and location of Dante's hell, turning the whole topic into a sort of geometrical problem.

Money was scarce, and Galileo had to earn his living. He did so by private tutoring, at which he was as good as Kepler was bad. But if he were to be financially secure, he needed a regular job, and he looked around to see what he could find. Rather surprisingly, he managed to obtain the chair of mathematics at Pisa University. This was in 1589, only four years after he had left the University without taking any degree at all. His first lecture was given on 12 November, and no doubt he felt a sense of deep satisfaction. After all, he had become a professor entirely on the strength of his own effort and ability.

Naturally, Galileo had to teach astronomy, and this meant accepting the Ptolemaic theory. Had he not done so, his career as a professor would have come to an abrupt end. Whether or not he actually believed in the Earth-centred system is a matter for debate. If he did, then it was not long before he changed his mind, and it seems likely that he became a convinced Copernican at a very early stage in his life, though it was not until some years later that he came out into the open even to his friends. Remember, he was still feeling his way, and even though he was blunt to the point of being aggressive he was much too sensible to court dismissal. So Galileo passed on to his students all the accepted Ptolemaic ideas—and, in private, went on thinking extremely hard.

It was during his three years at Pisa that he laid the foundations of some of the ideas about the science we now call experimental mechanics, and in particular he studied the motions of small balls rolling down inclined planes. This may not sound very important, but it led on to some notable developments such as the principle of uniform acceleration; and Galileo's faith in the teaching of Aristotle became less and less. This seems to be the moment to describe the famous experiment involving the Leaning Tower of Pisa, because it is significant even if it were not actually tried.

The Leaning Tower remains one of the most famous landmarks in Pisa, and it really is tilted strikingly from the vertical. According to the popular story, Galileo went to the top of it and released two weights, one much heavier than the other. On the old theories—known as Aristotelian, though it must be added that Aristotle's surviving books are rather vague about this particular principle—the heavier weight would hit the ground first. Galileo maintained that they would land at the same moment, because they would be accelerated to the same extent. Of course he was right, and the admiring crowd dispersed, thoroughly convinced.

Unfortunately there is no evidence that any special public demonstration ever took place. Viviani, whose biography of Galileo was finished only a dozen years after the great man's death, says that he dropped weights off the Tower not once, but many times. There is no reason to doubt him, but Viviani did not regard it as important, and Galileo never mentioned it in any of his writings. Actually there was nothing new in it. In a book published in 1605 a Dutch scientist, Simon Stevin, mentioned that he and others had tried the experiment quite often. We must also remember that air-resistance has to be taken into account, which is why a feather falls much more slowly than a pencil. The experiment can be carried out properly only in an airless environment, as was demonstrated on television so long afterwards by Apollo astronauts standing on the surface of the Moon.

As time went by, Galileo increased his attacks on the ideas which had been put forward by Aristotle and which had been regarded with such reverence ever since. For instance: 'Aristotle, in practically everything that he wrote about local motion, wrote the opposite of the truth.' And he was 'weary and ashamed of having to use so many words to refute such childish arguments and such inept attempts at subtleties as those which Aristotle crams in the whole book of *De Cœlo*.' Elsewhere: 'Aristotle was ignorant not only of the deeper and more abstruse discoveries of geometry, but even of its most elementary principles.' This would hardly endear Galileo to his fellow professors, who accepted Aristotle without question. If Aristotle had been ignorant, then so were they—with much less excuse.

In other directions, too, Galileo showed a signal lack of tact. There was his poetic lampoon *Against the Toga*, aimed at the costumes of his colleagues. His conclusion was that the toga should be discarded, and it would even be best to go around without wearing any clothes at all, though he did concede that this would be rather too extreme. There was also the time when the Grand Duke of Tuscany, Ferdinand I—who had actually granted him his professorship at Pisa—asked him to inspect a model of a dredging machine which had been designed by the Governor of the province of Leghorn, who also happened to be the Grand Duke's half-brother. It did not take Galileo long to see that the machine was fundamentally unsound. It would have been wise to compliment the inventor on his ingenuity, and then make some soothing comments to the effect that some drastic modifications might be needed. Galileo, being Galileo, said bluntly that the whole apparatus was futile, and a waste of time. The Governor went ahead, and had the machine built. When it was tested out in Leghorn harbour, it signally failed to work. The Governor was not pleased, and, human nature being what it is, some of his wrath was directed toward Galileo. The result was that

Galileo lost favour with the Grand Duke just at a time when he could ill afford to do so.

The situation at Pisa was getting worse and worse. Quite apart from the unfriendliness of the professors, there was the matter of money. Galileo's father Vincenzio had died in 1591, and this meant that someone had to look after his family; clearly this was the responsibility of the eldest son. The Pisa salary was not nearly enough, and after the unfortunate episode of the dredging machine there was little hope of an increase; indeed, there were doubts as to whether his contract would be renewed at all. So Galileo looked anxiously around, and applied for the chair of mathematics at the University of Padua, which had been vacant for some time. Padua was financed by the authorities of the Venetian Republic, and so in the autumn of 1592 Galileo went to Venice, where he was well received. Within a few weeks he had packed up his belongings, left Pisa (Leaning Tower and all) and installed himself in his new home. As he admitted later, the years he spent at Padua, between 1592 and 1610, were the happiest of his whole life, and had he stayed there instead of moving back to Florence he might have avoided at least some of his misfortunes. In outlook, the Venetian Republic was much more liberal than Tuscany.

Family troubles continued. Galileo had already had to provide a marriage dowry for one of his sisters, Virginia, and during the Paduan period he also had to do so for another, Livia. Michelangelo, his younger brother, proved to be a broken reed; he was no mean musician, but was incapable of holding down any professional post, and time and time again Galileo had to provide money for him and his children. Galileo himself never married, but for more than ten years he kept up an association with one Marina Gamba, by whom he had two daughters and one son. The association came to an end when Galileo left Padua in 1610, but apparently the two remained on excellent terms, and Galileo carried on a cordial correspondence with the man whom Marina eventually did marry.

Yet in spite of all these anxieties, and the need to supplement his income by private teaching, Galileo was able to enjoy himself as well as carrying on his research. He moved in circles which were both influential and intellectual; Padua is close to Venice, so that he was able to spend a good deal of time there too. One of his Venetian friends, Giovanfrancesco Sagredo, was to have his name immortalized in Galileo's most famous book long afterwards. Galileo was also well-known enough to be consulted on all sorts of matters. For instance, there was the time when a shipyard owner asked him whether it would be better to position the rowlock on the side of the boat or on a projecting strut. Galileo replied that since the oar acted as a lever, with the fulcrum at the blade, it did not in the least matter. Which pattern was adopted we do not know.

I do not propose to say much about Galileo's researches into mechanics made during his stay in Padua, because they are not relevant to my theme; suffice to say that they were many and varied. He set up a small workshop in his house, and made various mathematical instruments, such as a sort of calculating ruler known as a geometric compass. He even wrote a book about this compass, and took effective legal action against one Baldessar Capra, who had claimed priority for the invention – though it should be added that Capra had attacked Galileo's views on a previous occasion, so that there was no love lost between the two. Actually, much of the scientific work which Galileo carried out during these years was not published until near the end of his life, after the Rome trial, when he was officially a prisoner in his Arcetri villa.

I must, I think, add something about Galileo's researches into the problem of free fall, because they are of vital importance. Aristotle had followed the idea that 'everything must find its natural place', and there were two directions: up (air, fire) and down (earth, water). Light a match, and the flame points upward; drop a piece of mud, and it will fall to the ground. Galileo, on the other hand, believed that everything must tend to move toward the centre of the Earth, and that any upward motion is due to the body being surrounded by a denser medium which, so to speak, pushes it up just as a cork will rise to the top of a tub of water. In a letter to his friend Paolo Sarpi, written in 1604, Galileo stated plainly that 'a body in natural motion increases its speed in the same proportion as its departure from the origin of the motion' according to a definite relationship. In other words: if you drop a stone, it will fall a certain distance in the first second, a greater distance in the next second, and so on; it will be constantly accelerated. This is nothing more nor less than Newtonian gravitation, though Galileo made the mistake of supposing that the acceleration would stop as soon as the body had reached the 'proper' speed characteristic of it. He was undoubtedly on the right track, as is shown by his significant comment: 'Uniformly accelerated motion I call that to which, commencing from rest, equal velocities are added in equal times.'

With this aside, let us return to Galileo's life at Padua. He was professor of mathematics there, just as he had been at Pisa, and he still had to continue with official classes in the Ptolemaic theory. He discussed

Opposite: *Galileo Galilei 1563–1642 : portrait by Ottario Leoni circa 1624*

it quite conventionally, and his students departed with no outlandish ideas. His *Treatise on the Sphere,* written in 1597, is essentially Ptolemaic, though it does contain the comment that 'there have not been lacking very great philosophers and mathematicians who, deeming the Earth to be a star, have endowed it with motion'. Yet by this time there is absolutely no doubt that Galileo had become a convinced Copernican. He said so in a letter dated 30 May 1597, written to his friend Professor Mazzoni at Pisa, and also, of course, in his reply to Kepler after having received a copy of the *Mysterium.* He did not take up Kepler's invitation to teach Copernican ideas openly, but there was nothing surprising in this. He was paid to give lectures about the Ptolemaic universe, and he carried out his duties. Moreover, he was well aware that he could give no hard and fast proofs that Copernicus was right and Ptolemy wrong. We must not forget that the original Copernican system, with its epicycles and deferents, was hopelessly incorrect, and in pre-telescopic times it was very difficult to provide anything in the way of experimental evidence either way.

The year 1604 was notable for the appearance of Kepler's Star, the supernova in Ophiuchus. Galileo observed it, of course, and gave three public lectures about it. For reasons which are not entirely clear (and can never be made clear, because we do not have his lecture texts) he considered that the new star showed the fallacy of some of Ptolemy's arguments, and this was indeed the first time that he publicly came out on the side of Copernicus. His views were promptly criticized by Baldessar Capra; but Galileo made no reply then, though he was quick enough to deal with Capra later over the compass controversy.

Life at Padua went pleasantly on, but Galileo was not completely satisfied; his first home had been in Tuscany, and he had spent some time in restoring his favour with the Grand Duke. In particular he had tutored his son, Cosimo, later to succeed his father as the Grand Duke Cosimo II. Whether Galileo would have made a move from Padua but for his sudden leap to fame is something that will never be known. Meantime, events taking place in far-away Holland were setting the scene for the start of the final battle between the supporters of the two world systems. The telescope had been invented; and it was the telescope which altered Galileo's whole life.

Opposite: *Chandelier in Pisa Cathedral*

Telescopes and the Sky

There has been a great deal of discussion about the invention of the telescope. There is no doubt that spectacle lenses to help people with weak eyesight were in use from the late thirteenth century onward, and by the time when Galileo was growing up spectacles had begun to look rather like ours, made up of lenses held in a frame which rests on the nose. Some sort of instrument can be made out of a couple of suitable spectacle lenses, and it seems curious that telescopes were so long delayed. Recent research by C. A. Ronan seems to have shown definitely that a strange sort of telescope, involving both a lens and a mirror, was made by that shadowy figure, Leonard Digges, at some time between 1550 and 1560, but there is no proof that it was ever turned toward the sky. Later, in Holland, a Middleburg optician named Zacharias Jansen has been said to have made a two-lens combination which magnified distant objects, abandoning it because the image appeared upside-down. However, the first telescopes whose existence is beyond any doubt were made in the first decade of the seventeenth century by Jän Lippershey, a lens-grinder of Middleburg in the Dutch island of Walcheren. On 2 October 1608 Lippershey offered a telescope to the authorities in Holland, and was asked to make a two-eyed version, so that he proceeded to make binoculars – and sold three sets of them at high price. They gave erect images, and were intended for use in looking at terrestrial objects, ranging from ships out at sea to distant peaks. Yet there is a note in a brochure, dated 22 November 1608, that a telescope could also be used for "seeing stars which are not ordinarily in view, because of their smaliness".

Opposite: *An early Italian telescope*

Right: *Telescope used by Galileo and Torricelli*

The candidate usually favoured is Jan Lippershey, a lens-grinder who
lived at Middelburg on the Dutch island of Walcheren; the authority
quoted is Pierre Borel, who wrote a book in 1655 called *De vero telescopii*

Hans Lippershey. From Pierre Borel's De Vero Telescopii Inventore, *1655*

inventore and stated firmly that the credit must go to Lippershey. Other writers are not so sure. There is no doubt that spectacle lenses to help people with weak eyesight were in use from the late thirteenth century onward, and by the time when Galileo was growing up spectacles had begun to look rather like ours, made up of lenses held in a frame which rests on the nose. To me, the main problem has always been: Why was the invention of the telescope so long delayed? Some sort of instrument can be made out of a couple of suitable spectacle lenses, and these were available. It all seems very peculiar.

Various other claims have been made. There are vague and totally unconfirmed reports that Roger Bacon used a telescope as early as 1260, but I think these can be disregarded. We can also dismiss an Italian named della Porta, who wrote a famous book about magic in which he included some comments about lenses which were later misinterpreted. There is more to be said in favour of two Englishmen, Dee and Digges, but—as I have already noted—both these are elusive figures about whom we know depressingly little. In Holland, another Middelburg optician named Zacharias Janssen was once said to have made a two-lens combination which magnified distant objects, but to have abandoned it because the image appeared upside-down. It would take me too far afield to pursue the matter further here, but I am anxious to stress that whoever invented the principle of the telescope, it was most definitely not Galileo. We do know that on 2 October 1608 Jan Lippershey offered a telescope to the authorities in Holland, and was asked to produce a two-eyed version, so that he proceeded to make binoculars—and sold three sets of them at a high price. By 1609 telescopes were actually on sale in Paris, and presumably elsewhere. They gave erect images, and were intended for use in looking at terrestrial objects, ranging from ships out at sea to distant peaks. Yet there is a note in a brochure, dated 22 November 1608, that a telescope could also be used for 'seeing stars which are not ordinarily in view, because of their smallness'.

Neither was Galileo the first man to turn a telescope toward the sky, and I must put in an aside here, because I played a minor and quite undistinguished rôle in the publication of the first of all telescopic maps of the Moon. In 1965 I had a letter from Dr. E. Strout, of the Institute of the History of Science in the U.S.S.R., enclosing a paper which he had written about a lunar chart drawn up in July 1609 by Thomas Harriott, one-time tutor to Sir Walter Raleigh. Harriott, who was born in 1560 and died in 1621, was a skilled mathematician who actually took his degree at Oxford, and evidently he managed to obtain one of the very early telescopes; his manuscripts were in the care of the Earl of Egremont, which is were Dr. Strout had located them. I was then Director of the Lunar Section of the British Astronomical Association, and I at once arranged for the paper to be published in the Association's Journal, where it duly appeared (Volume 75, page 102). Harriott's chart was remarkably accurate, even though he was always reluctant to publish any of his scientific work and, so far as we know, never followed it up. Another pre-Galilean observer of the Moon was a Welshman, Sir William Lower, who compared the lunar surface to a tart that his cook had made: 'here some bright stuff, there some dark, and so confusedly all over'. But Harriott, Lower and the various other telescopic pioneers of the time were mere dabblers compared with Galileo, and the fact that the great Italian was not first in the field is neither here nor there.

Galileo first heard about the telescope in the summer of 1609, and set to work to make one for himself. His account, published in his book *Sidereus Nuncius*, tells us just what happened.

Two of Galileo's original telescopes, and the broken object-lens, preserved in the Tribuna de Galileo, Florence

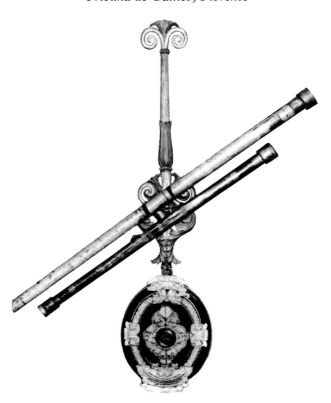

'About ten months ago a report reached my ears that a Dutchman had constructed a telescope, by the aid of which visible objects, although a great distance from the eye of the observer, were seen distinctly as if near; and some proofs of its most wonderful performances were reported, which some gave credence to, but others contradicted. A few days after, I received confirmation of the report in a letter written from Paris by a noble Frenchman, Jaques Badovere, which finally determined me to give myself up first to inquire into the principle of the telescope, and then to consider the means by which I might compass the invention of a similar instrument, which after a little while I succeeded in doing, through deep study of the theory of refraction; and I prepared a tube, at first of lead, in the ends of which I fitted two glass lenses, both plane on one side, but on the other side one spherically convex, and the other concave. Then bringing my eye to the concave lens I saw objects satisfactorily large and near, for they appeared one-third of the distance off and nine times larger than when they are seen with the natural eye alone. I shortly afterwards constructed another telescope with more nicety, which magnified objects more than sixty times. At length, by sparing neither labour nor expense, I succeeded in constructing for myself an instrument so superior that objects seen through it appear magnified nearly a thousand times, and more than thirty times nearer than if viewed by the natural powers of sight alone.'

This extract is enough to dispose of the unfair accusation that Galileo tried to pass off the invention as his own. Still, he made justifiable capital out of it, and in August 1609 he offered a telescope to the Venetian authorities, who were suitably enthusiastic. He was at once offered a permanent position at Padua University, together with a considerable increase in salary; and though he was still anxious to return to Tuscany, he felt bound to accept. In late August he wrote a letter to his brother-in-law:

'As the news had reached Venice . . . I was summoned before their Highnesses, and exhibited a telescope to them, to the astonishment of the whole Senate. Many of the nobles and senators, although of a great age, mounted more than once to the top of the highest church tower in Venice in order to see sails and ships that were so far off that it was two hours before they were seen without my spyglass, steering full-sail into the harbour; for the effect of my instrument is such that it makes an object fifty miles off appear as large as if it were only five.'

During his lifetime Galileo seems to have made around a hundred telescopes, of which a few survive. Each had a convex object-glass and a concave eyepiece, giving a right-way-up or erect image but a very small field. It was, ironically, Kepler who introduced a convex eyepiece, which improved the field and performance at the cost of inverting the image. Kepler did not actually make a telescope, so far as we know, and the first instrument on the 'Keplerian' pattern was not constructed until about 1615, by Christoph Scheiner, a Jesuit lecturer at the University of Ingolstädt; but Kepler's famous book *Dioptrice* shows that as a theoretical optician he was ahead of Galileo. But for the moment the telescopes made by Galileo were supreme; nobody else could hope to match them.

In the winter of 1609–10 Galileo set to work, making observations energetically and recording everything that he saw. Possibly no other period of discovery has been so fruitful, and the results were even more

Modern photograph of a half-moon

far-reaching than Galileo could have imagined when he started. It will, I think, be best to give an account of the observations first, with Galileo's own comments about them, and to go on to a discussion of how they were received by his colleagues – and his enemies.

The discoveries came in rapid succession. One of the things that Galileo found was that although the planets were magnified into disks, the stars were not: a telescope 'powerful enough to enlarge other objects a hundred times will scarcely render the stars magnified four or five times'. Galileo described how large numbers of stars, much too faint to be seen with the naked eye, came within his range: 'Beyond the stars of the sixth magnitude you will behold through the telescope a host

Galileo's map of the moon from his Sidereus Nuncius, *1610. Compare this with the photograph on the facing page*

of other stars, which escape the unassisted sight, so numerous as to be almost beyond belief. . . . In order that you may see one or two proofs of the inconceivable manner in which they are crowded together, I have determined to make out a case against two star-clusters, that from them as a specimen you may decide about the rest. As my first example, I had determined to depict the entire constellation of Orion, but I was so overwhelmed by the vast quantity of stars and by want of time that I have deferred attempting this to another occasion, for there are adjacent to, or scattered among, the old stars more than 500 new stars within the limits of one or two degrees. For this reason I have selected the three stars in Orion's Belt and the six in his Sword, which have been long well-known groups, and I have added eighty other stars. . . . As a second example I have depicted the six stars in the constellation Taurus, called the Pleiades (I say six intentionally, since the seventh is scarcely ever visible). . . . Near these lie more than forty others invisible to the naked eye, no one of which is more than half a degree off any of the aforesaid six; of these I have noticed only thirty-six in my diagram.' (The Pleiades are nicknamed the *Seven* Sisters. I once conducted an experiment in which I was able to show that anyone with average eyes can see seven stars in the Pleiades when the sky is dark and clear, but Galileo's remark can be explained on the basis that one of the seven, Pleione, is distinctly variable, so that in the early seventeenth century it may have been slightly fainter than it is now. People with exceptional sight can see well over a dozen stars in the Pleiades without optical aid, and the record is said to be nineteen.)

There was also the question of the Milky Way, which had been the subject of so many legends. Galileo wrote: 'By the aid of a telescope, anyone may see that all the disputes which have tormented philosophers through so many ages are resolved at once by the unmistakable evidence of our eyes, and we are freed from wordy disputes upon this subject, for the Milky Way is nothing else but a collection of innumerable stars planted together in groups. Upon whatever part of it you direct the telescope, straightway a vast crowd of stars presents itself to view; many of them are tolerably large and extremely bright, but the number of smaller ones is quite beyond determination.'

This was a most significant discovery. Galileo could not know that the Milky Way effect is due to our looking along the main plane of the Galaxy, but the sheer number of stars may have stressed yet again the real unimportance of the Earth in the universe.

When Galileo turned his telescope to the Moon (probably as soon as the instrument was complete) he could see a mass of detail. A few pre-

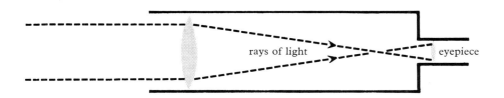

rays of light

eyepiece

Principle of the Galilean refractor.

telescopic maps had been drawn, and the Greeks had known the Moon to be 'earthy, with mountains and valleys'; there were also many legends about the Man in the Moon, and it is true that with an effort of the imagination one can make a picture out of the arrangement of light areas and darker plains. The dark regions, still mis-called seas, are less reflective and less uneven than the highlands; most of them are bordered by high mountain ranges, though these are admittedly rather different in nature from our own mountain chains such as the Himalayas. Thus the Mare Imbrium or Sea of Showers, the best-defined of the seas, has a border made up in part of the mountains which we now call the lunar Apennines and Alps.

Galileo wrote: 'Let me first speak of the surface of the Moon which is turned toward us. For the sake of being understood more easily, I distinguish two parts in it, which I call respectively the brighter and the darker. The brighter part seems to surround and pervade the whole hemisphere; but the darker part, like a sort of cloud, discolours the Moon's surface and makes it appear covered with spots. Now these spots, as they are somewhat dark and of considerable size, are plain to everyone, and every age has seen them, wherefore I shall call them *great* or *ancient* spots, to distinguish them from other spots, smaller in size, but so thickly scattered that they sprinkle the whole surface of the Moon, but especially the brighter portion of it.' (These are, of course, the craters.) 'These spots have never been observed by anyone before me; and from my observations of them, often repeated, I have been led to the opinion which I have expressed, namely, that I feel sure that the surface of the Moon is not perfectly smooth, free from inequalities and exactly spherical, as a large body of philosophers considers with regard to the Moon and other heavenly bodies, but that, on the contrary, it is full of inequalities, uneven, full of hollows and protuberances, just like the surface of the Earth itself, which is varied everywhere by lofty mountains and deep valleys.'

Of course it was well established that the Moon shines only by reflected sunlight, and it occurred to Galileo that this might give him a

155

means of measuring the heights of the peaks. 'Again, not only are the boundaries of shadow and light in the Moon seen to be uneven and wavy, but still more astonishingly many bright spots appear within the darkened portion of the Moon, completely separated and divided from the illuminated part and at a considerable distance from it. After a time these gradually increase in size and brightness, and an hour or two later they become joined up with the rest of the lighted part which has now increased in size. Meanwhile, more and more peaks shoot up, one here and another there. . . . Now, is it not the case on the Earth before sunrise, that while the level plain is still in shadow, the peaks of the most lofty mountains are illuminated by the Sun's rays? After a little while does not the light spread further, while the middle and larger parts of those mountains are becoming illuminated; and at length, when the Sun has risen, do not the illuminated parts of the plains and hills join together? The grandeur, however, of such prominences and depressions on the Moon seems to surpass both in magnitude and extent the ruggedness of the Earth's surface, as I will hereafter show.'

For his first measurements of lunar heights, Galileo used the shadow method. If you know the angle at which the sunlight is striking, and you also know the length of the shadow cast by the lunar mountain, it is a simple problem in geometry to work out the height of the mountain itself (Fig 12). Galileo deduced that the peaks were of the order of four or five miles high – as lofty as anything on the Earth, and much higher relatively, in view of the fact that the Moon is a smaller world. Actually he concentrated upon the Apennines, and his values were over-estimated, because even the highest of the Apennines rises to less than 18,000 feet; but under the circumstances the result was a good one. It was this kind of experiment which made Galileo stand head and shoulders above his contemporaries in telescopic astronomy. He wanted

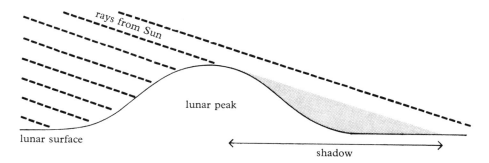

Figure 12: Galileo's method of measuring lunar altitudes.

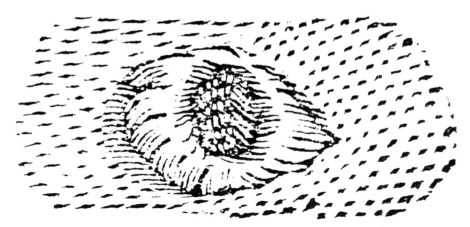

Lunar craters drawn by Galileo in 1611, published in Operere di Galileo
Galilei, *1655* ·

to draw conclusions as well as to make observations – and to publish
them. He made a few drawings of the Moon, and some of the features
are recognizable (notably the Mare Imbrium and the Apennine range),
though it cannot be denied that the slightly earlier map by Harriott was
both more detailed and more accurate.

It was natural to suppose that the bright regions were lands and the
dark patches water, and this opinion was held for some time, though it
had to be discarded as soon as more powerful telescopes showed masses
of detail on the dark plains also. From the samples brought back from
the Moon by the Apollo astronauts and the automatic Russian probes
we may now be sure that the 'seas' were never water-covered, but in
the earlier days of the Moon's existence they were probably filled with
liquid lava, so that the names are not entirely inappropriate.

Yet it was with the planets that Galileo was most concerned. By now
he had dedicated himself to spreading the Copernican system – Sun in
the middle, Earth in motion – but he was only too well aware that he
would need positive proof before he could hope to make any real impact
upon the orthodox ideas, which had been unchallenged ever since the
time of Aristotle. Copernicus himself had been unable to provide any
proof at all, and in any case his original system was very wide of the
mark. Galileo, with his continued belief in perfectly circular paths, was
also at fault; but at least he now had a means of putting some of the
basic ideas to the test. And it was with special interest that he turned
his telescope toward the brilliant Venus.

According to Ptolemy, both Venus and the Sun move round the Earth,
with Venus being the nearer of the two. In the diagram (Fig 13a) – which

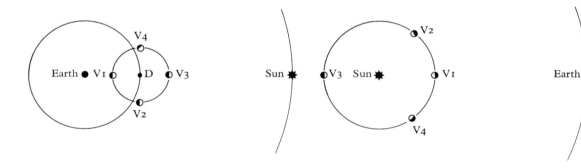

Figure 13: *left* Ptolemaic, *right* actual phases of Venus.

is, admittedly, very over-simplified–E represents the Earth and s the Sun; D is the deferent of Venus, and v1, v2, v3 and v4 show Venus itself in four positions of its epicycle. The line EDS must always be straight. Now, since the Sun can light up only half a planet at a time, it is quite obvious that if Venus moves in this way it can never be seen as a full disk, or even a half. At v1 and v3 it will be invisible altogether, because its dark side will be turned toward us; at other times it will appear as a crescent.

This was not what Galileo found. He showed a full range, from new (invisible) to practically full; and on any heliocentric theory–circular or-bits or not–this is perfectly logical. In the next diagram (Fig 13b) we again have the Earth and the Sun, but this time the Sun is in the centre, with Venus moving round it. It is new (and cannot be seen) at v1, full at v3, and half-phase at v2 and v4, so that it is gibbous, or between half and full, to either side of v3. When visible in the evening sky it is waning; when it shines forth in the eastern sky before dawn it is waxing.

There could be no doubt whatsoever that the phases of Venus knocked away one of the main pillars of the Ptolemaic system. Galileo first announced his discovery in anagram form: 'Hæc immatura a me iam frustra leguntur o.y.', or, in English, 'These things which are unready for disclosure are read by me'–the final letters o.y. being tacked on merely to make up the anagram. Rearranged, the letters give: 'Cynthia figuras æmulatur Mater Amorum' (the phases of Cynthia are imitated by the Mother of Love)–Cynthia being the Moon and the Mother of Love being Venus. At that period it was quite usual for a discovery to be given in anagram form, to establish priority, but there was no real need for it here, partly because there could be no question that Galileo had been the first to record the phases and secondly because they are so

Jupiter, in blue light, showing the Great Red Spot, Satellite Ganymede and shadow (above)

obvious. (Any modern binoculars will show them, and there are even a few keen-eyed people who can see them without any optical aid at all, though I would be the last to claim that I can do so myself.)

The other shattering blow to the Ptolemaic theory – actually the first, in chronological order of the observations – was provided by the discovery that Jupiter is attended by four satellites or moons of its own. Again I quote Galileo's own words, because it would be impossible to improve upon them:

'There remains the matter, which seems to me to deserve to be considered the most important in this work, namely, that I should disclose and publish to the world the occasion of discovering and observing four planets, never seen from the very beginning of the world up to our own times, their positions, and the observations made during the last two months about their movements and their changes of magnitude; and I summon all astronomers to apply themselves to examine and determine their periodic times, which it has not been permitted me to achieve up to this day, owing to the restriction of my time. I give them warning, however, again, so that they may not approach such an inquiry to no purpose, that they will want a very accurate telescope, and such as I have described in the beginning of this account.

'On the 7th day of January in the present year, 1610, in the first hour of the following night, when I was viewing the constellations of the heavens through a telescope, the planet Jupiter presented itself to my view, and as I had prepared for myself a very excellent instrument, I noticed a circumstance which I had never been able to notice before, namely, that three little stars, small but very bright, were near the planet; and although I believed them to belong to the number of fixed stars, yet they made me somewhat wonder, because they seemed to be arranged exactly in a straight line, parallel to the ecliptic, and to be brighter than the rest of the stars, equal to them in magnitude. The position of them with reference to one another and to Jupiter was as follows:

'On the east side there were two stars, and a single one toward the west. The star which was furthest toward the east, and the western star, appeared rather larger than the third.

'I scarcely troubled at all about the distance between them and Jupiter, for, as I have already said, at first I believed them to be fixed stars; but when on 8 January, led by some fatality, I turned again to look at the same part of the heavens, I found a very different state of things, for there were three little stars all west of Jupiter, and nearer together than on the previous night, and they were separated from one another by equal intervals, as the accompanying figure shows:

'At this point, although I had not turned my thoughts at all upon the approximation of the stars to one another, yet my surprise began to be excited, how Jupiter could one day be found to the east of all the aforesaid fixed stars when the day before it had been west of two of them; and forthwith I became afraid lest the planet might have moved differently from the calculation of astronomers, and so had passed those stars by its own proper motion. I, therefore, waited for the next night with the most intense longing, but I was disappointed of my hope, for the sky was covered with clouds in every direction.

'But on 10 January the stars appeared in the following position with regard to Jupiter, the third, as I thought, being hidden by the planet:

They were situated just as before, exactly in the same straight line with Jupiter, and along the Zodiac . . .

'When I had seen these phenomena, I knew that corresponding changes of position could not by any means belong to Jupiter, and as, moreover, I perceived that the stars which I saw had always been the same, for there were no others either in front or behind, within a great distance, along the Zodiac–at length, changing from doubt into surprise, I discovered that the interchange of position which I saw belonged not to Jupiter, but to the stars to which my attention had been drawn, and I thought therefore that they ought to be observed hence-forward with more attention and precision.

'Accordingly, on 11 January I saw the arrangement of the following kind:

East   ★   ★   ☆   West

namely, only two stars to the east of Jupiter, the nearer of which was distant from Jupiter three times as far as from the star further to the east; and the star furthest to the east was nearly twice as large as the other one; whereas on the previous night they had appeared nearly of equal magnitude. I therefore concluded, and decided unhesitatingly, that there are three stars in the heavens moving about Jupiter, as Venus and Mercury round the Sun; which at length was established as clear as daylight by numerous other subsequent observations. These observations also established that there are not only three, but four, erratic sidereal bodies performing their revolutions round Jupiter . . .

'These are my observations upon the four Medicean planets, recently discovered for the first time by me; and although it is not yet permitted me to deduce by calculation from these observations the orbits of these bodies, yet I may be allowed to make some statements, based upon them, well worthy of attention.

'And, in the first place, since they are sometimes behind, sometimes before Jupiter, at like distances, and withdraw from this planet toward the east and toward the west only within very narrow limits of divergence, and since they accompany this planet alike when its motion is retrograde and direct, it can be a matter of doubt to no-one that they perform their revolutions about this planet, while at the same time they all accomplish together orbits of twelve years' length about the centre of the world. Moreover, they revolve in unequal circles, which is evidently the conclusion to be drawn from the fact that I have never been permitted to see two satellites in conjunction when their distance from Jupiter was great, whereas near Jupiter two, three, and sometimes all four, have been found closely packed together. . . . We have a notable and splendid argument to remove the scruples of those who can tolerate the revolution of the planets round the Sun in the Copernican system, yet are so disturbed by the motion of the Moon about the Earth, while both accomplish an orbit of one year's length about the Sun, that they consider that this theory of the universe must be upset as impossible; for now we have not one planet only revolving about another, while both traverse a vast orbit about the Sun, but our sense of sight presents to us four satellites circling about Jupiter, like the Moon about the Earth, while the whole system travels over a mighty orbit about the Sun in the space of twelve years.'

The vital point here was that by his observations, Galileo had established that there was more than one centre of motion in the universe. The 'Medicean planets'—as he called them, for reasons which will become clear shortly—moved not round the Earth, or for that matter round the Sun, but round Jupiter. This was in absolute contradiction to the old belief that the Earth must be the centre of everything, and Galileo was quick to realize the importance of what he had found.

The account given here is a translation from Galileo's book the *Sidereus Nuncius (Starry Message)*, which led on to the first of the great conflicts with the Church. The book was dedicated to Galileo's former pupil, now the Grand Duke Cosimo II of Tuscany, whose family name was de'Medici. Despite his successes at Padua, Galileo still wanted to return to Tuscany, and he needed to retain the favour of the Grand Duke. He proposed to name the four new bodies Catharina or Franciscus, Maria or Ferdinandus, Cosmus Major and Cosmus Minor. Not surprisingly, these names were badly received, and neither did astronomers outside Italy like the term 'Medicean planets'. Different names were put forward by Simon Mayr (better known as Marius), another telescopic observer, and it is his names which we use today: Io, Europa, Ganymede and Callisto, all of which are conventionally mythological.

There ensued a rather undignified squabble between Galileo and Marius with regard to who had first recognized the four satellites. Certainly Marius was using a telescope quite as early as Galileo, and perhaps earlier; he may well have seen the satellites, but whether he recognized them as being different from ordinary stars is quite another matter. Had he done so, it is reasonable to think that he would have made an announcement, particularly as he was on friendly terms with his countryman Kepler. He would surely have told Kepler; but he did not—and Kepler first heard of the discovery from another friend, who rejoiced in the name of Wackher von Wackenfels.

Interestingly, it seems definite that during his early observations of the Jovian satellites, Galileo actually recorded the planet Neptune! He sketched its position once in December, 1612 and twice in January 1614, though naturally he mistook it for a star.

In July 1610 Galileo, using his most powerful telescope, managed to see the disk of the planet Saturn. The magnification was only 32, and the definition given by the telescope was poor, but Galileo realized that there was something very peculiar about Saturn's shape. In a letter to Vinta, secretary of state to the Grand Duke Cosimo, he wrote that 'the planet Saturn is not one alone, but is composed of three, which almost touch one another and never move nor change with respect to each other. They are arranged in a line parallel with the Zodiac, and the middle one is about three times the size of the lateral ones.' As usual, he issued an anagram, made up of a jumble of letters; Kepler mis-interpreted it, and spent some time searching fruitlessly for two moons not of Saturn, but of Mars.

Galileo was puzzled by Saturn. The two attendant bodies were quite unlike the satellites of Jupiter, and showed no relative motion at all. Two years later, in 1612, he was even more astonished to find that the planet appeared single; and he wrote: 'What is to be said concerning so strange a metamorphosis? Are the two lesser stars consumed after the manner of the spots of the Sun? Have they vanished or suddenly fled? Has Saturn, perhaps, devoured his own children? Or were the appear-ances indeed illusion or fraud, with which the glasses have so long deceived me, as well as many others to whom I have shown them? . . . I do not know what to say in a case so surprising, so unlooked-for, and so novel.' Later the secondary bodies appeared again, and began to look rather like arms or handles attached to the main disk.

Galileo was not alone in being baffled by Saturn, and some very peculiar theories flourished for some decades after his discovery. Even in the 1650s a French mathematician named Gilles de Roberval was still maintaining that Saturn must be surrounded by a torrid zone giving off vapours, transparent if in small quantity, reflecting sunlight at the edges if of medium density, and producing an elongated elliptical aspect if very thick, while another Frenchman, Honoré Fabri, was led to suggest that there were several satellites, some bright and some dark. The problem was eventually solved by Christiaan Huygens, in 1656, who announced (in an anagram, needless to say!) that Saturn is surrounded by a 'flat ring which nowhere touches the body of the planet'. When the rings are edge-on to the Earth they cannot be seen at all in small or moderate telescopes, and this was what happened in 1612, when the Earth passed through the ring-plane as shown in the diagram. (It is often said that the rings vanish completely as seen through any telescope. I cannot agree, because during the edge-on presentation in 1966 I was able to keep them in constant view, using the 10-inch refractor at the Armagh Observatory.) It is not surprising that the 'secondary bodies' vanished from Galileo's sight.

Yet another controversy centred around the discovery of spots on the Sun. In fact, naked-eye sunspots had been recorded now and then all through scientific history, but nobody had ever been able to understand what they could be. There is no doubt that the first observer to publish telescopic observations of them was Johannes Fabricius, a young Dutchman (son of David Fabricius) whose book came out in 1611, but which – infuriatingly – gives no dates for the original drawings. Probably Fabricius saw sunspots for the first time at the end of 1610. Scheiner at Ingolstadt, recorded spots in March 1611, with his pupil C. B. Cysat. His discovery was not well received by his ecclesiastical superior, Busaeus, who said simply: 'I have read all the works of Aristotle several times from beginning to end, and I assure you that I have not found anything in them which could be what you are telling me. Go, my son, and calm yourself. I assure you that what you took to be spots on the Sun are only flaws in your glasses or in your eyes.'

Scheiner was not satisfied; and when he had reports of similar observations by others he wrote some letters to his friend Mark Welser, in Augsburg, who had them printed. A copy of the tract reached Galileo, who replied to the effect that he had been observing sunspots ever since November 1610. If this is so, then he was the first to identify them.

At any rate Galileo gave a correct explanation, whereas Scheiner did not. Remember, Scheiner did not (officially) believe in the Copernican theory, and he was disinclined to believe that a body as pure as the Sun could possibly be spotted. So he preferred the idea that the spots were dark bodies close to the Sun, whereas Galileo was able to follow their movements as they were carried across the solar disk from day to day – and this led on to the discovery that the Sun takes between three and four weeks to spin once on its axis.

Galileo's description of sunspots sounds remarkably modern in tone. 'The dark spots seen on the solar disk by means of the telescope are not at all distant from its surface, but are either touching it or are separated by an interval so small as to be quite imperceptible. Nor are they stars or other permanent bodies, as some are always produced and others dissolved. They vary in duration from one or two days to thirty or forty. For the most part they are of irregular shape, and their forms continually change, some quickly and violently, others more slowly and moderately. They also vary in darkness, appearing sometimes to condense and sometimes to spread out and rarefy. In addition to changing shape, some of them divide into three or four, and often several unite into one.'

This is exactly how sunspots do behave. They are not truly dark, as Galileo realized; but their temperatures are two thousand degrees or so below that of the surrounding bright surface or photosphere, and they appear blackish by contrast.

There is a story that Galileo ruined his eyesight by looking at the Sun through a telescope, and that this led to his becoming blind during the last years of his life. I am frankly dubious, because in a letter to Welser he described the method he used for observing the sunspots— and it is the same as the method that any sensible person will use today. 'Direct the telescope upon the Sun as if you were going to observe that body. Having focused and steadied it, expose a flat white sheet of paper about a foot from the concave lens; upon this will fall a circular image of the Sun's disk, with all the spots that are on it arranged and disposed with exactly the same symmetry as in the Sun.' The method, apparently, was first thought out by Galileo's pupil Benedetto Castelli. Of course, Galileo may have made some preliminary observations by 'looking direct' and injured his eyes in the process; but on the whole I doubt it. Certainly he would have been blinded at once if he had turned one of his more powerful telescopes sunward and then looked through the eyepiece. (Kepler records that he once turned his telescope toward the Sun, 'with its upper lens open no wider than the head of a pin. And here I am, although an hour has already passed, with every single letter that I am writing fiery red. My eyes burned and itched painfully, even though I protected myself with dark glass, and endured the Sun's brilliance for scarcely the twinkling of an eye.')

Galileo's book *Letters on Sunspots,* in which he described his observations, caused a storm because it was openly pro-Copernican; and it must be said that Father Christoph Scheiner played a highly discreditable part in later proceedings. But the first of the books in which the telescopic discoveries were given was the *Sidereus Nuncius,* and with its publication, in March 1610, came the start of what was to prove the decisive battle in the scientific revolution.

The Storm Gathers

The *Sidereus Nuncius* was a short book, but it was one of the most important in the history of science, and unlike Copernicus' *De Revolutionibus* its impact was immediate. As we have seen, it was dedicated to the Grand Duke Cosimo II, and it had precisely the effect which Galileo had hoped. Within a few weeks of publication Galileo had had an invitation to become 'Chief Mathematician of the University of Pisa, and Philosopher to the Grand Duke', without any teaching obligations, and at a handsome salary. He accepted without demur, and moved to Florence in September 1610 – to the intense annoyance of the Venetian authorities, who had only just confirmed his permanent appointment at Padua and who had expected him to stay where he was. In retrospect, Galileo's move was a sad mistake. He would have done far better to stay in the more liberal atmosphere of Venice.

There was a personal complication as well. Marina Gamba did not go with Galileo, and so far as we can tell they never met again. Their young son Vincenzio remained with his mother for a time, while the two daughters entered a convent near Florence, apparently rather against their wishes. Both became nuns, and it was the elder, Sister Maria Celeste, who proved such a comfort to Galileo after the Rome trial of over twenty years afterwards.

Up to now Galileo had done nothing to bring the full fury of the Church down on his head. True, he had made it known that he believed the Sun to be the centre of the Solar System, but in his official lectures at Padua he had taught the conventional Ptolemaic theory, as he was required to do by the terms of his contract. But his telescopic discoveries altered the whole situation, if only because they could not possibly be

Cardinal Robert Bellarmine, a portrait in the Uffizi Gallery, Florence

reconciled with the idea of a central and motionless Earth. Also, nobody who read the *Sidereus Nuncius* could be in any doubt as to what Galileo really thought.

One man who did read it, and who agreed wholeheartedly with it, was Johannes Kepler, who received a copy in April 1610 (before Galileo left Padua). With the book was a request for Kepler's opinion, and Kepler made haste to comply. His lengthy letter was completed in a tremendous hurry, and was subsequently published. It was called *Conversation with the Sidereal Messenger*, a title which led to some unfortunate repercussions. The Latin word 'nuncius' (or 'nuntius') may be taken to mean either 'message' or 'messenger'. Galileo evidently intended the former; Kepler took him to mean the latter – and unwittingly gave Galileo's enemies the chance to claim that Galileo had tried to pass himself off as a messenger from heaven. To avoid further confusion, I propose now to call the book simply the *Nuncius*.

Kepler's *Conversation* is to all intents and purposes a reasoned and logical discussion of the various discoveries that Galileo had made, and the theories that he had put forward. Remember, Kepler had as yet no telescope, even though he had published an excellent work on optics, and clearly he wanted one. In his book he wrote: 'With eagerness, then, I await your instrument. . . . Yet if Fate smiles on me so that I can overcome the obstacles and attempt the mechanical construction, I shall exert myself energetically in that endeavour, pursuing alternative courses.' He then goes on to describe the improved form of telescope, which in fact he never personally built. Neither did Galileo take the hint and send Kepler a telescope of his own, which under the circumstances was rude as well as unwise. Yet Galileo was busy sending telescopes to people whom he regarded as influential, and one of these was dispatched to the Elector of Cologne, who passed it on to Kepler.

Most of the *Nuncius* deals with observational matters, and Kepler's opinions here are of interest. He agreed with Galileo that the dark patches on the Moon were sheets of liquid, while the bright areas were land; he believed that the Sun was much more powerful than the stars, so that it represented an unique body in the universe; and he was suitably enthusiastic about the discovery of the four satellites of Jupiter. He had a few criticisms, but all of these were minor. Under the

Opposite: *The Leaning Tower of Pisa*

Overleaf: *The Pleiades star cluster : 48-inch Schmidt photograph from the Mount Wilson and Palomar observatory*

circumstances it is surprising that Galileo did not reply; but he did nothing, and once more the correspondence between the two great men lapsed.

Kepler did take care to observe the satellites of Jupiter for himself as soon as he had obtained a telescope, and in 1611 he published his account of them in a book generally known as the *Narratio* or *Narrative*. Other people were less co-operative, and the limit was reached by a professor at Pisa, Julius Libri, who disbelieved everything that Galileo had said but who stubbornly declined to look through a telescope at all. When Libri died, in late 1610, Galileo is reported to have said that he might at least have a good view of the Jovian moons as he passed by them on his way to heaven.

The storm-clouds were starting to gather, but at first they seemed very distant and unimportant. At Padua, Galileo's old friend Cesare Cremonini made it clear that he had no time for telescopes or such devices, and when asked about Galileo's discoveries made the classic retort: 'I believe that nobody but Galileo has seen them; and besides, looking through those spectacles gives me a headache. Enough; I do not want to know any more about it.' Actually there is no evidence that Cremonini ever used a telescope, but his objections were not personal ones, and he remained on terms of friendship with Galileo despite the tremendous scientific rift between them.

Various adverse comments found their way into print—for instance, one by Martin Horky in 1610, *A Very Brief Excursion Against the Sidereal Messenger*—and there were some powerful enemies at Bologna, headed by one Giovanni Magini, who did try to use a telescope which Galileo took him, though apparently with scant success. Father Christopher Clavius, a very well-known professor of mathematics at the Collegio Romano, also thought that the 'discoveries' were no more than faults in Galileo's lenses, though, to his credit, he later changed his mind, and even said that the whole picture of the universe might have to be rearranged. It was Clavius, incidentally, who tried to reconcile Galileo's observations of the Moon with the old Aristotelian theory that the lunar globe must be smooth and perfect. He assumed that the Moon's mountains and valleys were covered with a completely transparent crystalline layer, so that the globe was indeed smooth even though it looked rough! Galileo replied caustically that this was 'a beautiful flight of the imagination'.

Opposite: *The object lens with which Galileo discovered the four satellites of Jupiter. The lens, now broken is mounted in an ivory frame.*

Meantime, Galileo had decided that, all things considered, it would be a good idea to pay a visit to Rome. By now he was wholly committed to spreading what he believed to be the truth – that the Sun, not the Earth, was the centre of the universe; and he genuinely thought that by means of his telescopes, plus his own powers of persuasion, he could overcome all opposition not only on scientific grounds, but on religious ones too. The Church did not seem to be unfriendly; Galileo was on excellent terms with people such as the powerful Cardinal Maffeo Barberini; and there was no reason to expect opposition from the Pope, Paul V. So with the full approval of Duke Cosimo of Tuscany, Galileo set out for Rome.

He arrived in the early part of 1611, and his reception was just as good as he had hoped. The cardinals and the Pope treated him with the greatest courtesy, and he had long discussions with men such as Clavius, who was now fully reconciled to the fact that the satellites of Jupiter really existed.

There were, of course, a few ominous signs, though Galileo may have been unaware of them. The influential Jesuit cardinal Robert Bellarmine went to the Collegio Romano to ask Clavius and other authorities there just what they thought about Galileo's work. Bellarmine was in something of a quandary. He had used a telescope, and he was too sensible to

Original letter of Galileo to Kepler, 4 August 1597, after having received a copy of Kepler's Mysterium Cosmographicum *(Cosmographic Mystery)*

deny the reality of what he saw, but he was suspicious of the implications, and clearly he thought that he must keep a watching brief on the turn of events. Above all, he was anxious to make sure that any new views of the universe would not conflict with religious teaching; echoes of Luther's old cry about 'the fool' who wanted to turn astronomy upside-down. But Bellarmine was given nothing definite to guide him. Clavius and the other mathematicians told him the ways in which they agreed with Galileo, and also the criticisms which they could produce. The Holy Office in Rome was kept informed, but nothing of importance happened, and Galileo returned home, no doubt highly satisfied with the result of his visit.

There was still considerable opposition. For instance, the authorities at the University of Pisa expressly forbade any new professors to teach that the Earth might move, or even to discuss the possibility among themselves. Also at Pisa the professor of philosophy, Boscaglia, declared that although Galileo's discoveries were real enough, it was sheer heresy to suggest that the Sun might be the centre of the universe. There was also Lodovico delle Colombe, who believed wholeheartedly in Aristotle and was rigorously opposed to Copernicanism. At one debate, held in the presence of the Grand Duke Cosimo, Galileo and Colombe had a heated argument, which I mention here because another participant was none other than Cardinal Barberini—and he supported Galileo. I must add, however, that on this occasion the main argument was not about the alleged motion of the Earth, but about the behaviour of bodies floating in liquid, which is outside my main theme.

Yet the storm-clouds did not disperse, and the first real attack on Galileo on religious grounds was made in October 1612. It was due to a Dominican, Nicolò Lorini, who showed such remarkable ignorance that Galileo was amused rather than annoyed; Lorini actually wrote to him a few days later, making what could be regarded as a lame apology. Much more serious was a sermon delivered in December 1614, in Florence, by another Dominican named Tommaso Caccini. Caccini attacked not only Galileo, but also Copernicus and every other mathematician, going so far as to claim that all scientists of this type represented a danger to the Church and hence to the State as well. At this point Lorini re-entered the fray, and sent the Holy Office a copy of a letter which, he thought, might prove incriminating.

The letter had been written by Galileo to Benedetto Castelli some time earlier (during 1613). It dealt with the connection between science and religion, and arose from a discussion between Castelli and some of the distinguished professors at Pisa University. Galileo was a staunch

Catholic, but he was becoming aware that his revolutionary views might lead to trouble with at least some Church authorities, and it seemed time to make his attitude clear. In his letter, he wrote: 'I think it would be the better part of wisdom not to allow anyone to apply passages of Scripture in such a way as to force them to support, as true, conclusions concerning Nature, the contrary of which may afterwards be revealed by the evidence of our senses.' In other words, the Bible need not be taken literally. Entirely without malice (in fact, quite the reverse) Castelli had shown the letter around, and various copies of it had been made. It was one of these copies which had fallen into Lorini's hands, and he made the most of it.

The Holy Office showed some alarm, and the Archbishop of Pisa was instructed to obtain the original copy of Galileo's letter; the version held by Lorini was carefully studied to see whether or not it could be regarded as heretical. The opinion given by an independent committee was rather vague, but not unfavourable to Galileo on the whole, and with any reasonable luck the affair would have gone no further. But Caccini was not to be put off, and early in 1615 he went to the Holy Office, demanding to be allowed to testify against Galileo. All sorts of charges were made, but in essence they came to the same thing. Galileo was preaching heresy, and would have to be silenced.

By this time Galileo was starting to realize that matters were becoming really serious. He managed to recover the original letter which had caused the trouble, and in 1615 he issued a revised and expanded version of it, entitled 'On the Use of Biblical Quotations in Matters of Science'. It was cleverly written, and Galileo hoped that it would prove persuasive enough to stifle criticism once and for all. As he pointed out, there is very little astronomy in the Bible, and of the planets only Venus is mentioned at all.

There were various Churchmen who took the same line. Among them was Paolo Foscarini, who wrote a booklet about it and sent his work to Cardinal Bellarmine for an opinion. Bellarmine's reply was so significant that it must be quoted. 'It appears to me that your Reverence' (i.e. Foscarini) 'and Signor Galileo did prudently to content yourselves with speaking hypothetically and not positively, as I have always believed Copernicus did. For to say that assuming the Earth moves and the Sun stands still saves all the appearances' (to save the appearances, or save the phenomena means to make the theory fit in with the observed data) 'better than eccentrics and epicycles is to speak well. This has no danger in it, and it suffices for mathematicians. But to wish to affirm that the Sun is really fixed in the centre of the heavens, and merely turns

Pope Paul V, detail from a portrait in the Pinacoteca Vaticana, Rome

upon itself without travelling from east to west, and that the Earth is situated in the third sphere and revolves very swiftly about the Sun, is a very dangerous thing, not only because it irritates all the theologians and scholastic philosophers, but also because it injures our holy faith and makes the sacred Scriptures false.'

Bellarmine's view was very similar to that of the modern Fundamentalist. Nobody in their senses could doubt that the Copernican theory, or some modification of it (remember that Kepler's first two Laws had been published by this time) was an invaluable mathematical tool, but Bellarmine simply could not bring himself to take it literally.

It was at this point that Galileo made up his mind to go back to Rome. His last visit had been a triumph; Barberini was still his friend; there seemed no reason to doubt that Pope Paul V would be as courteous as before. And yet, with the wisdom of hindsight, it seems that he was wrong to make the journey. The Florentine ambassador, Guicciardini, was more perceptive; he knew that Galileo had not stayed long enough in 1611 for the forces of opposition to be fully mustered against him, and that by now Rome was 'no place to argue about the Moon, nor to support nor import any new doctrines'. Galileo may even have been somewhat naïve about it all. His scientific authority was very consider-able, and possibly it did not occur to him that the Church would be sufficiently bigoted and ill-informed as to order him to desist from teaching or writing. The key figure was Robert Bellarmine, and when Galileo reached Rome he hoped for an interview.

The interview took place – but not in the way that Galileo had expected.

At some time during February 1616, Pope Paul decided to intervene personally. So far as he was concerned, the whole affair had gone far enough, and it was time to put an end to what he had come to regard as a tiresome controversy. Naturally, he asked for Bellarmine's advice, and Bellarmine said that Galileo's teachings were, in all probability, contrary to the Bible. Others held the same view. To make sure that there could be no misunderstanding, the Holy Office drew up two propositions and sent them to the leading scholars for a firm decision as to whether they were dangerous or not. The propositions were:

1. That the Sun is the centre of the universe: and
2. That the Earth is neither motionless nor the centre of the universe, but revolves round the Sun, as well as rotating on its axis once a day.

It took the scholars less than a week to reply. They judged that the first proposition was foolish and philosophically absurd as well as being heretical, while the second was almost as bad. The die was cast, and so far as the Holy Office was concerned there could be no turning back.

On 25 February, the Pope ordered Cardinal Bellarmine to summon Galileo and tell him to stop spreading these dangerous doctrines. A day later the interview took place. It must have been a strange meeting, and a bitter one for Galileo, if only because it was in such stark contrast to anything which had happened before. There was no appeal; had Galileo refused to comply he would have been imprisoned at best, tortured at worst (and it was only sixteen years since Giordano Bruno had been burned at the stake, in the same city). There is some argument as to what Bellarmine actually said, and the wording became important later, as we will see, but the main message was clear. Galileo must refrain from defending the false doctrine of a spinning Earth moving round a stationary Sun.

The next step was to put all the relevant writings on the Index of Prohibited Books. This was duly done. Among them was Foscarini's tract, while *De Revolutionibus* was banned 'pending certain corrections' (which, of course, were never made and could not be made). They remained on the Index for more than two hundred years; the first list which did not contain them was that of 1835, during the girlhood of

Queen Victoria, though the decision to free them had been taken more than a decade earlier.

It sounds incredible that even in the time of Napoleon, no devout Catholic was allowed to read a classic scientific book written in the sixteenth century by a Polish canon, but once again we must not be too ready to laugh. As recently as the 1960s the teaching of Darwin's theory of natural selection was still banned from schools in some parts of the United States. Human nature is slow to change.

Undoubtedly the 1616 visit to Rome was as disastrous for Galileo as his earlier trip had been triumphal, but at least he was not in strong personal disfavour. When he saw the Pope he was given a cordial welcome, and he also had a second audience with Cardinal Bellarmine, at which he received a written statement to the effect that he had done nothing wrong except to teach theories which had now been exposed as baseless and heretical. In a way this may have reacted to his later disadvantage, because it could have led him on to expect better treatment from the Church than he actually received.

For the moment he could do no more. Headstrong though he was, he knew better than to go against a direct order from the notorious Holy Office; the Inquisition was not to be trifled with. So in June 1616 Galileo left Rome and went back to Florence, as convinced a Copernican as ever, but resigned to postponing his great campaign and biding his time. No doubt the situation would alter; meanwhile he could at least continue with his observation and research, so that he could present an even more impressive case for the movement of the Earth as soon as circumstances allowed him to do so.

It must have been irksome, none the less. For instance, early in 1616 Galileo had been sent a long letter from a clerical scholar named Ingoli, giving what he regarded as scientific disproofs of the Copernican system and drawing heavily upon Aristotle, Ptolemy and Tycho Brahe. It would have been easy enough for Galileo to dispose of everything that Ingoli said. Now, alas, he was not free to do so; there was no option but to keep silent, whatever the provocation.

Yet even though Copernicanism could not be discussed openly, there were other subjects which were still unrestricted, and an opportunity came in 1618, when three bright comets appeared in rapid succession. This was most unusual, and has not happened since, though admittedly there have been years graced by two impressive cometary visitors (1910, for instance). By bad luck Galileo was unable to observe any of the 1618 comets, because he was ill in bed with a severe attack of arthritis, but at least he could voice his opinions – and he did.

The three comets caused wide interest, not unmixed with alarm. (The fear of comets is still not quite dead. When Kohoutek's Comet showed indications of becoming spectacular at the beginning of 1974, I received stacks of booklets from religious eccentrics who forecast all manner of disasters, from floods up to volcanic explosions and even the end of civilization. The fact that the comet did not make the display it had been expected to do is neither here nor there!) Dozens of booklets and pamphlets were published about them, among which Galileo selected a completely harmless lecture delivered by Orazio Grassi, professor of mathematics in the Collegio Romano. Grassi was not a Copernican; he followed Tycho Brahe's system of the universe, and his lecture said nothing new. As with Tycho and the comet of 1577, he noted the lack of detectable parallax, and stated that the comet must certainly be farther away than the Moon.

This was logical, and Galileo might have been expected to agree, but nothing was further from his mind, and we have to admit that his opinions about comets were completely wrong—much more erroneous than Grassi's or, for that matter, Tycho's. The trouble, so far as he was concerned, was the rather obvious fact that a comet did not move in a circular path either round the Earth or round the Sun. Its motion was quite different, and did not appear to fit in with any accepted system at all.

Aristotle had explained comets to his own satisfaction as being hot, dry exhalations, rising from the ground and carried along by the motion of the sky. When they were sufficiently heated by the motion, they caught fire, and either burned up rapidly as shooting-stars or else more slowly to produce comets, fanned by the warm breezes from the lower atmosphere. Galileo did not accept this theory—one can hardly imagine him taking anything that Aristotle had said at face value!—but his own idea was no better; he regarded a comet as being simply the effect of the refraction of sunlight in vapours rising from the Earth, so that it would not be expected to show parallax any more than a rainbow does. A comet did not fit into the Copernican pattern, and so could not be 'real'.

It is not easy to see how a man with Galileo's penetrating and analytical mind could make so complete a miscalculation. The very appearance of a comet shows that it is something more than a mere optical phenomenon. But there it was; and in attacking Grassi, Galileo also made strong criticisms of Tycho Brahe. In fact Grassi's original tract had been anonymous, and Galileo's came out under the ostensible authorship of his pupil Mario Guiducci; it was called *Discourse on the Comets*.

The Comet of 249 A.D. *Woodcut from Lycosthenes'* Prodigiorum ac Ostentorium
Chronicon, *Basle, 1557*

Nobody was under the slightest doubt that Galileo was the real author, and when Grassi published a reply, later in 1619, he did so under the name of 'Sarsi'. In retrospect it all seems rather childish, and there is little point in going through the various arguments about the nature of comets, because Galileo's theory was so very wide of the mark. But one comment of Grassi's in his *Libra* of 1619 is worth recording. From beneath his cloak of anonymity, he wrote: 'Let it be granted that my master followed Tycho. Is this such a crime? Whom instead should he follow? Ptolemy, whose followers' throats are threatened by the outthrust sword of Mars now made closer? Copernicus? But he who is pious will rather call everyone away from him, and will spurn and reject his recently-condemned hypothesis. Therefore Tycho remains as the only one of whom we may approve as our leader among the unknown courses of the stars.'

There is some justification in claiming that Grassi was hitting below the metaphorical belt, because he knew quite well that the 1616 Papal decree meant that Galileo could not possibly leap to Copernicus' defence. Yet the matter was not to rest there, and Galileo set to work on another book, published in 1623 under the title of *Il Saggiatore* or *The Assayer*. It contained some faulty science, but it was so skilfully written – in Italian, by the way, not Latin – that it was widely acclaimed.

Galileo did not restrict himself to discussing comets; *Il Saggiatore* covered a wide range, extending from astronomy to physics, mechanics and philosophy. It also showed up the folly of arguing from tradition instead of experience. Moreover, there were notes on various practical experiments that Galileo had made in order to prove his points. For instance, Aristotle had believed that vapours rising from the Earth were carried round by the motion of the sky; Galileo claimed that nothing so light as air could be swept along by touching the surface of its container – and he showed what was meant by placing a lighted candle in the centre of a hollow vessel, revolving the vessel and observing that the candle-flame remained erect, so that obviously the air was not moving. He was right, too, in saying that friction could cause heat, and he took Grassi to task for his rather uncritical acceptance of some other old ideas – such as the belief that cannon-balls sometimes melted in mid-air because of their speed. Galileo wrote, devastatingly: 'If Sarsi wishes me to believe . . . that the Babylonians cooked eggs by whirling them rapidly in slings, I will believe it; but I will say that the cause of the effect is very far from the one he attributes to it.'

By the time that *Il Saggiatore* appeared two other things had happened, each of which affected Galileo's personal career. In 1621 Cosimo II, Grand Duke of Tuscany and Galileo's old pupil, died unexpectedly and was succeeded by his son Ferdinand II, who was still a minor and had therefore to have a council of regency. Even when he came of age Ferdinand was weak and ineffectual, so that Galileo was deprived of powerful support. Pope Paul V died in the same year, and was succeeded by Gregory XV, about whom nothing need be said here because he held the papacy for only a brief period, left no mark on history, and died in 1623. Who was to succeed Gregory? The choice fell on no less a person than Cardinal Maffeo Barberini, who took the title of Urban VIII.

Galileo was overjoyed. Barberini was an old friend, with whom he had never had any differences; surely the political and religious climate must now change? *Il Saggiatore* was dedicated to the new Pope, who responded just as he might have been expected to do. Galileo went back to Rome in person and had half a dozen audiences with Urban, who was as friendly as ever. Yet there was a subtle difference. Cardinal Maffeo Barberini had been able to talk 'off the record', and to give his personal views even if they were not in absolute accord with official doctrine; Pope Urban VIII could not, as he made clear from the outset.

What Galileo wanted, of course, was a cancellation of the 1616 Decree which forbade him to teach or discuss the movement of the

Earth, but this was something about which the Pope was adamant. Urban was in a difficult position. Whether or not he really believed in the Copernican system is something about which we can never be sure, but he did his best to steer a middle course without making any concessions which would be regarded as risky. He even put forward what has become known as 'the argument of Urban VIII', to the effect that even though many phenomena made it appear that the Earth orbits the Sun, it was still quite possible that God, in His infinite wisdom, had produced these effects by making the Sun orbit the Earth just as was stated in official doctrine. Galileo was not impressed, but it was not easy to decide what to do next. If he were to continue his pro-Copernican crusade, he certainly had to act in some way or other; if he let the opportunity slip, it would be unlikely to recur.

It seemed to him that the attack by Francesco Ingoli would make a good test case. In 1616 Galileo had been unable to make any reply at all; now the time had come to do so, but he made it clear that he meant to deal with purely scientific and philosophical aspects, leaving theology to the theologians. He had faith in Urban, and it was logical to think that the Pope and other enlightened Churchmen would realize that nothing could stifle the progress of science indefinitely. So Galileo did his best to be conciliatory, and in his booklet he wrote, ostensibly to Ingoli: 'It remains for me to beg you to receive in good spirit these replies of mine, as I hope you will do, both through your innate courtesy and because it is thus that all lovers of truth should act.' There was little here of the fiery intolerance that had marked Galileo's younger days.

The booklet was quite widely read, and probably the Pope was among those who saw it. There were no strong protests from anyone, and Galileo felt encouraged. And yet the 1616 Decree was still in force, and *De Revolutionibus* was still on the Index 'pending corrections'; Galileo had enemies in Rome and elsewhere, and it would have been foolish to be over-confident. The only real answer was to write a book which would accomplish two things: describe the 'two systems of the world', Ptolemaic and Copernican, and also bring out the truth in such a way that the Papal decree would not be infringed. Galileo believed this to be possible, so in 1624 he set to work. Had he been able to see into the future, he might well have changed his plans.

DIALOGO
di
GALILEO GALILEI LINCEO
AL SER.mo FERD. II. GRAN. DVCA DI
TOSCANA

Stefan Della Bella F

The Dialogue

It was not to be expected that a lengthy book such as Galileo now planned could be written quickly. Not only had the text to be intelligible to the layman, but it also had to be presented in a way which would not cause trouble with the Church. It was a tremendous task, and by now Galileo was no longer a young man; he had passed the age of sixty, and he had never spared himself. Moreover he had outside troubles, inasmuch as his shiftless brother Michelangelo returned to Italy and calmly brought his wife and seven children to live with Galileo, where they remained for some time.

Yet nothing would stand in Galileo's way, and finally, either at the very end of 1629 or the first days of 1630, the great book was finished. He meant to call it *Dialogue on the Tides*, though the censors made him alter it to *Dialogue Concerning the Two Chief World Systems*. After numerous delays it appeared in print in February 1632. Before dealing with the events leading up to Galileo's trial and condemnation it is, however, essential to go into some detail about the book itself. It was originally published in Italian, but by now it has been translated into practically every language of mankind. (In particular there is the splendid English translation by Stillman Drake, published in 1953 with a foreword by Einstein. It is from this translation that I have taken all the quotes which follow.)

The book is a long one, and I do not propose to do more than give a few extracts, bringing out the points which seem to be most relevant to the main story. Let it be said at once that in my view, at least, the *Dialogue* is not a technical treatise in the accepted sense of the term. There is not much in it which had not been said before somewhere or other. It was a brilliant work of what we may call popularization, and it

Opposite: *Frontispiece of Galileo's* Dialogue, *Florence 1632. From left to right, the three figures represent Aristotle, Ptolemy and Copernicus*

summed up the situation with devastating clarity. Kepler could never have written in such a vein, even had the occasion demanded; neither could Copernicus; Tycho could have made an attempt, but without Galileo's wit and overall mental grasp the book would have been a certain failure.

His initial problem was to put his case forward without running foul of the 1616 Decree. To write a book in defence of Copernicanism, outlining the Ptolemaic theory and then demolishing it, was therefore out of the question. The book would have to be presented as an impartial discussion, and Galileo therefore introduced three characters, two of them real and the third (prudently) fictitious. The *dramatis personæ* were:

1. Filippo Salviati, a Florentine nobleman and an old friend of Galileo; indeed, the *Letters on Sunspots* had been dedicated to him. In Galileo's words, Salviati had 'a sublime intellect which fed no more hungrily upon any pleasure than it did upon fine meditations'. He had died in 1614 at the early age of thirty-one.

2. Giovanni Francesco (or Giovanfrancesco) Sagredo, 'a man of noble extraction and trenchant wit', also an early and close friend of Galileo's. The two had spent much time together in Venice, where Sagredo died in 1620.

3. Simplicio—no Christian name is ever given, and who was not ostensibly drawn from any living person. I say 'ostensibly' because it was the name and outlook of Simplicio which caused so much later trouble with the Pope.

In the *Dialogue*, the three friends meet together at Sagredo's palace in Venice to discuss the pattern of the universe, with special reference to the problem of whether the Earth travels round the Sun or whether it lies at rest in the very centre of the universe. Of course, the three have different outlooks. Salviati is strongly in favour of Copernicus, and has no patience whatsoever with either Aristotle or Ptolemy—or, for that matter, with Tycho Brahe. Simplicio is purely Aristotelian; he cannot believe that the Earth moves, and he accepts the old ideas without reservation. The rôle of Sagredo is that of an intelligent onlooker, who is willing to listen to both sides and who guides the conversation gently along whenever it shows signs of flagging. In modern parlance, we would describe Sagredo as a skilful question-master.

This sounds innocent enough; but is it? First, there is not the shadow of a doubt that Galileo identifies himself with Salviati, who is the dominant figure throughout the book, and who talks at least twice as much as anybody else. Salviati is not only far-sighted, but is also sharp-witted, and this makes Simplicio appear in a very poor light. Indeed, Simplicio is sometimes unable to grasp obvious facts even when pointed out to him in detail, and he tends to become confused; Salviati leads him along, now and then chiding him for his slowness on the uptake and his complete refusal to question anything that Aristotle had said. '. . . The power of truth is such that when you try to attack it, your very assaults reinforce and validate it.'

Therefore, it is the Copernican who is intelligent and the Aristotelian who is stupid. All that Simplicio can really do is to repeat the old doctrines, more or less parrot-fashion, and to go over the same ground time and time again in the manner of a modern politician. He can put forward no telling objections to the arguments set out by Salviati, and his own theories are knocked down with smooth efficiency. Nobody who reads the *Dialogue* through can pretend that it is impartial. The very name for Simplicio was, incidentally, a bad error of judgement on the part of Galileo—because it was promptly, even if unjustifiably, identified with no less a dignitary than Pope Urban VIII. . . .

The *Dialogue* is spread over four days. Broadly speaking, the first is devoted to the destruction of Aristotle's ideas, the second and third to proving the motion of the Earth on its axis and in orbit round the Sun, and the fourth to Galileo's theory of the tides, which was in fact completely erroneous even though he meant it to be the crux of the whole Copernican argument. As a start, Salviati sums up the old views. 'Aristotle . . . separates the whole [of the universe] into two differing and, in a way, contrary parts; namely the celestial and the elemental, the former being ingenerable, incorruptible, unalterable, impenetrable, etc.; the latter being exposed to continual alteration, mutation, etc.' Motion is of three kinds: circular, straight, and a mixture of straight and circular. Salviati (or, rather, Galileo!) then shows that Aristotle is inconsistent in drawing an apparently natural distinction between straight-line and circular motion. 'This is the cornerstone, basis and foundation of the entire structure of the Aristotelian universe. . . . Now whatever defects are seen in the foundations, it is reasonable to doubt everything else that is built upon them.' He goes on: 'It may be immediately concluded that if all integral bodies in the world are by nature movable, it is impossible that their motions should be straight, or anything else but circular; and the reason is very plain and obvious. For whatever moves straight changes place and, continuing to move, goes ever further from its starting-point and from every place through which it successively passes. If that were the motion which naturally suited it, then at the beginning it was not at its proper place. . . . Besides, straight motion being by nature infinite (because a straight line is infinite and indeterminate), it is impossible that anything should have by nature the principle of moving in a straight line; or, in other words, towards a place where it is impossible to arrive, there being no finite end. For nature, as Aristotle well says himself, never undertakes to do that which cannot be done, nor endeavours to move whither it is impossible to arrive.'

Simplicio has no answer to this sort of attack (indeed, during the first pages of the *Dialogue* he scarcely even manages the proverbial word-in-edgeways!) and Salviati goes on to the question of the shape of the Earth, which, he shows, is spherical. 'Now, just as all the parts of the Earth mutually co-operate to form its whole, from which it follows that they have equal tendencies to come together in order to unite in the best possible way and adapt themselves by taking a spherical shape, why may we not believe that the Sun, Moon and other world bodies are also round in shape merely by a concordant instinct and natural tendency of all their component parts? If at any time one of these parts were forcibly separated from the whole, is it not reasonable to believe that it would return spontaneously and by natural tendency?'

Simplicio is now ready to come to Aristotle's defence. '. . . It is vain to inquire, as you do, what a part of the globe of the Sun or Moon would do if separated from the whole, because what you inquire into would be the consequence of an impossibility. For, as Aristotle demonstrates, celestial bodies are invariant, impenetrable and unbreakable; hence such a case could never arise. And even if it should, and the separated part did return to the whole, it would not return thus because of being heavy or light, since Aristotle also proves that celestial bodies are neither heavy nor light.'

This passage sums up Simplicio's whole attitude. There is no need for scientific proof; Aristotle has spoken, and he cannot be wrong. Simplicio then rebukes Salviati: 'Please, Salviati, speak more respect-fully of Aristotle. He having been the first, only, and admirable expounder of the syllogistic forms, of proofs, of disproofs, of the manner of discovering sophisms and fallacies – in short, of all logic – how can you ever convince anyone that he would subsequently equivocate so seriously as to take for granted that which is in question?'

Simplicio is very insistent upon the division of the universe into two parts, the lower or changeable and the celestial or incorruptible. Celestial bodies are 'neither hot nor cold'; the Earth is supreme, and nothing ever changes in the heavens, if only because no alterations have ever been seen there. Salviati has a prompt reply. 'But if you have to content yourself with these visible, or rather these seen experiences, you must consider China and America celestial bodies, since you have never seen in them these alterations which you see in Italy. Therefore, in your sense, they must be inalterable.' The obvious next step is to cite the cases of the supernovæ or new stars of 1572 and 1604, 'which were indisputably beyond the planets'. Sunspots, too, indicate change – of course Simplicio repeats Scheiner's old theory that they are due to dark bodies near the Sun – and Sagredo makes a pungent comment: 'Simplicio is confused and perplexed, and I seem to hear him say, "Who would be there to settle our controversies if Aristotle were to be deposed?"'

When they discuss the Moon, Sagredo comments that the outer parts of it at least are subject to alteration, but Simplicio disagrees. Any such changes would be useless, 'because we plainly see and feel all generations, changes, etc. that occur on Earth are either directly or indirectly designed for the use, comfort and benefit of man. . . . Now of what use to the human race could generations ever be which might happen on the Moon or other planets?' This gives Salviati the chance to launch out into a detailed and accurate description of the lunar world, together with the cause of eclipses and of the so-called Earthshine, or light reflected on to the night side of the Moon from the Earth. Then comes a typical exchange between the two main adversaries:—

Simplicio . . . I consider the Moon's sphere to be as smooth and polished as a mirror, whereas that of this Earth which we touch with our hands is very rough and rugged. . . . I do not believe that the Moon is entirely

Proportional compass designed by Galileo, made by Marcantonio Mazzolen whom Galileo hired in Padua to make instruments he designed

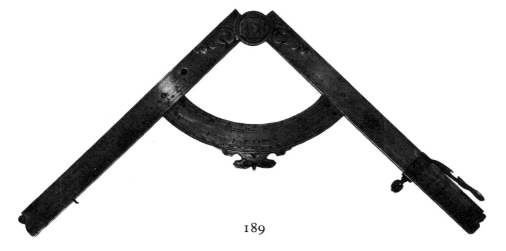

without light, like the Earth. On the contrary, that brightness which is observed on the balance of its disk outside of the thin horns lighted by the Sun I take to be its own natural light; not a reflection from the Earth, which is incapable of reflecting the Sun's rays by reason of its extreme roughness and darkness. . . . I concur in judging the body of the Moon to be very solid and hard like the Earth's. Even more so, for if from Aristotle we take it that the heavens are of impenetrable hardness and the stars are the denser parts of the heavens, then it must be that they are extremely solid and most impenetrable.

Sagredo What excellent stuff, the sky, for anyone who could get hold of it for building a palace! So hard, and yet so transparent!

Salviati Rather, what terrible stuff, being completely invisible because of its extreme transparency. One could not move about the rooms without grave danger of running into the doorposts and breaking one's head.

Certainly Galileo was a master of ridicule, but later in the first session of the *Dialogue* he is more blunt. Simplicio has repeated his claim that the Moon is as smooth as a mirror, and that 'the appearance you speak of, the mountains, rocks, ridges, valleys, etc., are all illusions.' Salviati returns to this, describing the progress of sunrise over the lunar peaks—which, it will be remembered, Galileo had used to make some height measurements. This time the *coup de grâce* is given by Sagredo: 'But please, Salviati, waste no time on this particular, because anyone who has had the patience to make observations of one or two lunations and is not satisfied with this very sensible truth could well be adjudged to have lost his wits.' Obviously this is directed at Simplicio, who answers lamely: 'Really, I have not made such observations, having neither the curiosity nor the instruments suitable for making them.'

The discussions go on, and each page is full of interest, but I have probably said enough to show the general trend of the book. The old theories are given, criticized and shown to be false. Blind faith in Aristotle is no substitute for scientific inquiry; and at the start of Day Two, Sagredo gives an amusing story to drive home the point still further. He relates how an anatomist had been dissecting a body, and had shown that the main nerves originate not in the heart, as Aristotle had taught, but in the brain. An Aristotelian philosopher who had been watching the experiment said, at length: 'You have made me see this matter so plainly and so palpably that if Aristotle's text were not contrary to it, stating clearly that the nerves originate in the heart, I should be forced to admit it to be true.'

Most of Day Two is taken up with proving that the Earth rotates on its axis, so that there is no need to believe that the sky turns around the Earth. Salviati puts forward cogent arguments. For instance: '. . . Let us consider only the immense bulk of the starry sphere in contrast with the smallness of the terrestrial globe, which is contained in the former so many millions of times. Now if we think of the velocity of motion required to make a complete rotation in a single day and night, I cannot persuade myself that anyone could be found who would think it the more reasonable and credible thing that it was the celestial sphere which did the turning, and the terrestrial globe which remained fixed.' And: 'Who is going to believe that Nature (which by general agreement does not act by means of many things when it can do so by means of few) has chosen to make an immense number of extremely large bodies move with inconceivable velocities, to achieve what could have been done by a moderate movement of one single body around its own centre?' Simplicio tries to counter-attack by using all the stock arguments, such as the claim that a rotating Earth would produce a constant wind from the east and that the overall effect of a rapid rotation would fling houses and trees high into the air. It is Sagredo who points out some of the fallacies here, adding, succinctly, that though many former followers of Aristotle and Ptolemy had turned Copernican, there was no contrary case. There are long sections in this part of the *Dialogue* about the movements and accelerations of various bodies, but then comes an unanswerable sally by (of course!) Salviati:

'Shut yourself up with some friend in the main cabin below decks on some large ship, and have with you some flies, butterflies and other small flying animals. Have a large bowl of water with some fish in it; hang up a bottle that empties drop by drop into a wide vessel beneath it. With the ship standing still, observe carefully how the little animals fly with equal speed to all sides of the cabin. The fish swim indifferently in all directions; the drops fall into the vessel beneath; and in throwing something to your friend, you need throw it no more strongly in one direction than another, the distances being equal; jumping with your feet together, you pass equal spaces in every direction. When you have observed all these things carefully (though there is no doubt that when the ship is standing still everything must happen in this way), have the ship proceed with any speed you like, so long as the motion is uniform and not fluctuating this way and that. You will discover not the least change in all the effects named, nor could you tell from any of them whether the ship were moving or standing still. . . . The cause of all these correspondences of effects is the fact that the ship's motion is common to all the things contained in it, and to the air also.'

Following a comment by Sagredo, Salviati delivers another thrust. 'I cannot help mentioning something I have noticed many times, and not without amusement. It occurs in nearly everyone who hears for the first time of the Earth's motion. Such people so firmly believe the Earth to be motionless that not only do they have no doubt of its being at rest, but they really believe that everyone else has always agreed with them in thinking it to have been created immovable and kept so in all past ages. Rooted in this idea, they are stupefied to hear that someone grants it to have motion, as if such a person, after having held it to be motionless, foolishly imagined it to have been set in motion when Pythagoras (or whoever it was) first said that it moved, and not before. Now that a silly idea like this, of supposing that those who admit the Earth's motion believe it first to have been stable, from its creation up to the time of Pythagoras, and then made movable only after Pythagoras deemed it to be so, should find a place in the giddy minds of common people is no marvel to me; but that the Aristotles and the Ptolemies should also have fallen into this puerility seems to be strange and inexcusable simple-mindedness.'

By now Sagredo is becoming convinced, and gives an example of how easy it is to be misled. 'The event is the appearance to those who travel along a street by night of being followed by the Moon, with steps equal to theirs, when they see it go gliding along the eaves of the roofs. There it looks to them just as would a cat really running along the tiles and putting them behind it; an appearance which, if reason did not intervene, would only too obviously deceive the senses.'

Simplicio tries another tack: no body which is not jointed can move in several senses at once, as Salviati believes the Earth to do. Pointedly, Salviati asks him: 'Are you being serious, or are you speaking ironically?' to which poor Simplicio can only reply 'I am giving you the very best that is in me.' At the end of Day Two Salviati gives another summary, and Simplicio has to admit: 'I have nothing else to say, except that the discussions held today certainly seem to me full of the most acute and ingenious ideas adduced on the Copernican side in support of the Earth's motion. But I do not feel entirely persuaded to believe them; for after all, the things which have been said prove nothing except that the reasons for the fixedness of the Earth are not necessarily reasons.'

With Day Three we come to the problem of the Earth's orbital movement round the Sun. The general approach is much the same as before; statements by Salviati, attempted rebuttals by Simplicio, and periodical interventions by Sagredo in his capacity of impartial referee. There is a long discussion about the status of the new star of 1572; a

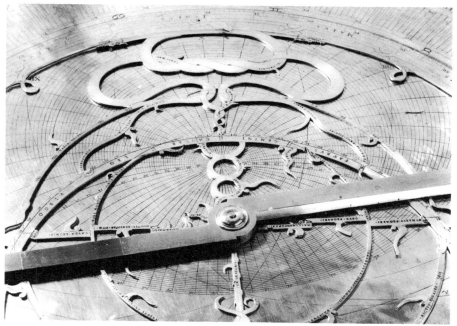

Astrolabe, made in Italy in the late sixteenth century. Although not designed by Galileo, there is some reason to believe that he used it while in Padua from 1592–1610

tract by a philosopher named Chiramonti had attempted to show that the star must be sublunar (i.e. closer than the Moon) and Galileo, in his rôle of Salviati, shows that this is not so. Simplicio makes few comments – there is little that he can say. But he returns to the fray when Salviati re-introduces the concept of a moving Earth, and, skilfully led by Salviati, he walks helplessly from one trap into another. He asks for observational proofs of the Earth's motion round the Sun, and is given them at once. 'The most palpable of these, which excludes the Earth from the centre and places the Sun there, is that we find all the planets closer to the Earth at one time and farther away from it at another. The differences are so great that Venus, for example, is six times as distant from us at its farthest as at its closest, and Mars nearly eight times as high in the one state as in the other. You may thus see whether Aristotle was not some trifle deceived in believing that they were always equally distant from us.' Next comes a description of why the outer planets (Mars, Jupiter and Saturn) pass through conjunction and opposition, and why Venus shows lunar-type phases. (Galileo did not doubt that the same must be true of Mercury, though his telescopes were not strong enough to give

193

confirmation.) When Sagredo wonders why so few people accept the Copernican system, Salviati answers: 'It seems to me that we can have little regard for imbeciles who take it as a conclusive proof in confirmation of the Earth's motionlessness, holding them firmly in this belief, when they observe that they cannot dine today at Constantinople or sup in Japan, or for those who are positive that the Earth is too heavy to climb up over the Sun and then fall headlong back down again. There is no need to bother about such men as these, whose name is legion, or to take notice of their fooleries. . . . Besides, with all the proofs in the world what would you expect to accomplish in the minds of people who are too stupid to recognize their own limitations?'

There are further discussions about the phases of Venus, the rôle of Jupiter's four satellites (which, as seen from Jupiter, would show phases similar to those of the Moon seen from Earth) and so on. Salviati again refers to the clumsiness of the Ptolemaic system, which could admittedly 'save the phenomena' and account more or less for the motions of the planets in the sky, but was hopelessly unsatisfactory. 'Thus however well the astronomer might be satisfied merely as a calculator, there was no satisfaction and peace for the astronomer as a scientist.' He shows that the apparent movements of sunspots as they pass across the brilliant disk, by virtue of the solar rotation, can also be fitted in with the theory that the Earth is travelling round the Sun.

Simplicio will not give up, but he is making no headway, as is shown by a comment from Sagredo: 'Simplicio is behaving bravely, and he battles very cleverly to sustain the Aristotelian and Ptolemaic side. To tell the truth, it seems to me that conversing with Salviati even for such a short time has considerably increased his capacity to reason rigorously.'

Another part of Day Three is devoted to the size of the universe, and Simplicio first introduces the subject, maintaining that on the Copernican pattern the great distances involved are 'incomprehensible and unbelievable'. Salviati takes him up, and shows that the opposition is mistaken in supposing that on Copernicus' theory a fixed star must be larger than the orbit of the Earth. He gives a method of showing that a star has an apparent diameter much less than that assumed by Tycho (about two minutes of arc). In Salviati's words: 'I hung up a light rope in the direction of a star (I made use of Vega, which rises between the north and the north-east), and then by approaching and retreating from this cord placed between me and the star, I found the point where its width just hid the star from me. This done, I found the distance of my eye from the cord, which amounts to the same thing as one of the sides which includes the angle formed at my eye and extending over the breadth of the cord. This is similar to, or rather equal to, the angle made in the stellar sphere by the diameter of the star. From the ratio of the thickness of the cord to the distance from my eye, using a table of arcs and chords, I immediately found the size of the angle. . . . By this very precise operation I find that the apparent diameter of a star of the first magnitude . . . is no more than five seconds'—or only a

Portrait of Galileo on the frontispiece of Systema Cosmicum, *Florence 1635*

very small fraction of the value given by Tycho, and commonly accepted up to that time. I have grave doubts as to whether Galileo ever tried this experiment, though he undoubtedly wanted his readers to believe that he did. While writing the present book, I tried it myself. I am no Galileo; but I found the experiment absolutely impossible to perform. Anyone who feels able to do better is more than welcome to try!

Salviati stresses that there is no need to suppose that the space between the orbit of Saturn and the fixed stars is empty; there could be many unseen bodies there, and he again brings out the fact that the Earth and other bodies are very tiny indeed compared with the vastness of the universe, so that we have no cause to assume that the stars would change in brilliance as the Earth swung toward and away from different parts of the starry sphere by virtue of its motion round the Sun. Before the end of Day Three there is a further concise summary of the Copernican system, together with some comments on the make-up of the Earth itself.

The fourth and last day of discussion is quite different. Galileo undoubtedly regarded it as the most important of all, but in fact it depends upon his completely erroneous theory of the tides—originally outlined in 1616 in a *Discourse on the Ebb and Flow of the Seas*. At that time the cause of tidal phenomena was not known; there was one idea, drawn from a casual remark in Aristotle's book about meteorology, that they were due partly to wind-force and partly to the natural slope of the Mediterranean, but other ideas had also been proposed, some of them linked with the Moon. Simplicio says that 'there are many who refer the tides to the Moon, saying that this has a particular dominion over the waters; lately a certain prelate has published a little tract wherein he says that the Moon, wandering through the sky, attracts and draws up toward itself a heap of water which goes along following it, so that the high sea is always in that part which lies 'under the Moon'. The 'certain prelate' was Marcantonio de Dominis. Salviati criticizes the idea strongly on the grounds that the water rises and falls only near the coasts, and not in the middle of the Mediterranean. Of course this is quite wrong, and de Dominis was working essentially along the right lines, so that for once Simplicio has the advantage. But Galileo did not realize this, and he maintained that no tides could ever occur on a motionless Earth.

His own theory was distinctly peculiar, and can best be explained by a diagram (Fig 14). The Earth is moving round the Sun, in the direction of the arrow, and is also spinning on its axis, again in the direction shown by an arrow. Consider any particular port—Venice, if you like—in its

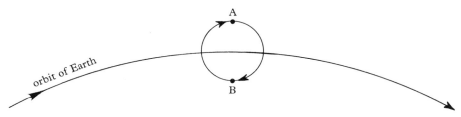

Figure 14: Galileo's tidal theory.

night-time position (A) and its daytime position (B), when it is turned sunward. At A, the orbital velocity and the axial spin add up, so that the waters are 'left behind' as the firm land rushes on. At B, the axial spin is opposed to the velocity in orbit, so that the sea 'catches up'. The result is a high tide every 24 hours, always around noon. This did not agree with observation, because at Venice there are two daily high tides rather than one, and they certainly do not occur regularly at noon. Galileo attributed this to the depth and profile of the Mediterranean.

I find it very hard to understand how Galileo, who placed such reliance upon observed facts, could have fallen into this trap, and how he could have reproached Kepler for the 'puerility' of believing that the tides could be affected by the Moon. The truth may be that he was desperately looking for some demonstration of the Earth's motion which nobody could deny, and the tides appeared to give it. Undoubtedly it was a 'blind spot' in his reasoning, though by now discerning readers were so used to finding Salviati right and Simplicio wrong that some of them presumably accepted it. Galileo himself gave up the theory from 1637, though he replaced it with another which was no better and which need not concern us here. It is ironical that the last day of the *Dialogue*, intended as a grand finale, should be the only really misleading part of the entire book.

In these few pages I have not attempted to give anything like a complete summary of Galileo's great work, and all I have tried to do is to present a general sketch of its content and its theme. I have not been deliberately selective in stressing the keen logic of Salviati at the expense of the traditional outlook of Simplicio; this was done by Galileo himself. It was necessary, but it was also dangerous—and the storm was about to break at last.

Confrontation!

Now let us return to the sequence of events.

Galileo finished the manuscript of the *Dialogue* within a few days of Christmas 1629, but obviously it was not simply a matter of sending it to press. Everything had to pass through the hands of the Papal censors, and Galileo was much too wise to attempt to by-pass the official procedure (even if this had been possible, which it was not). However, there seemed no cause for alarm. His old friend Benedetto Castelli had been appointed mathematician to the Pope, and in March a letter from him arrived, assuring Galileo of the friendly attitude of several people. These included the well-known scholar Federico Cesi; the Tuscan ambassador in Rome, the Marquis Francesco Niccolini; and, more important still, Niccolò Riccardi, who was in effect chief censor to the Vatican. Moreover, there was not the slightest reason to doubt that Urban VIII was still very much on Galileo's side – as he had been in 1616, when as Cardinal Barberini he had voiced his opposition to the notorious decree which had silenced Galileo for so long.

There were, of course, enemies as well, mainly among the Jesuits. Probably the most dangerous of them was Father Christoph Scheiner, who had disputed with Galileo the priority in the matter of sunspots. Since then Scheiner had made regular observations of the Sun, and he eventually produced a book with the strange title of *Rosa Ursina*, containing some valuable research. Unfortunately it also contained an attack on Galileo, for whom Scheiner evidently had a strong personal dislike. Galileo made no reply, but the episode was somewhat ominous.

Meantime, everything continued according to plan. Near the end of March 1630 Galileo made yet another journey to Rome, bringing the manuscript with him; it seemed only sensible to present it personally

Opposite: *Urban VIII (Cardinal Maffeo Barberini) portrait in Galleria Capitolina, Rome*

and ask for the approval of the Church authorities. He was well received, and saw both Niccolò Riccardi and another official, Rafael Visconti, who had been given the task of editing the book if necessary. There was no hint of trouble; Galileo was told that only a few minor modifications would be called for, and in June he was back in Florence, armed with full authorization to begin negotiations with a printer.

What did Riccardi really think? One can hardly believe that he could take the *Dialogue* to be an impartial work; unless he were signally lacking in intelligence, he must have realized that it was nothing more nor less than pro-Copernican propaganda. On the other hand, both sides of the case had been put forward, and presumably it was not the author's fault if the arguments given by Salviati were more convincing than those of Simplicio. At any rate, Riccardi was satisfied for the moment, and the book was duly sent back to Galileo.

One of Galileo's staunch allies, Cesi, died suddenly in the summer, and this caused a change in the printing arrangements, with which Cesi had been concerned. On the advice of Benedetto Castelli, Galileo then looked round for a printer in Florence, and found one without difficulty (after all, was he not official mathematician to the Grand Duke Ferdinand II of Tuscany?). But before the actual work could begin, Riccardi in Rome started to have some second thoughts.

Illustration from Scheine's Rosa Ursina *1630*

It is not certain why he became uneasy. Perhaps some of the Jesuits, notably Christoph Scheiner, were doing their best to make trouble; it is also possible that Riccardi, thinking about what he had read, decided that it might not be so innocent as it had looked. In any case, he instructed Galileo to return the manuscript so that it could be re-checked. This was the last thing that Galileo wanted to do, partly because he had no wish to make any further alterations and partly because he was anxious to speed up publication as much as possible. He did persuade Riccardi to let the Florentine censor, Giacinto Stefani, deal with the main text; but he was forced to send the preface and the conclusion back to Rome.

Stefani completed his task quickly, and apparently had no objection to anything in the book, but Riccardi adopted delaying tactics, and for a long time nothing happened. The year 1630 passed by, and so did the first part of 1631, despite regular appeals by Galileo and the Tuscan Ambassador. It was only in midsummer that Riccardi gave his final decision. Some changes were made, and any mention of the tides was to be dropped from the title, presumably because the tidal theory was regarded by Salviati as the all-important proof that Copernicus was right and Ptolemy wrong. Galileo made no demur. At last, it seemed, the way was clear; the censors had approved everything that he had written, and the book could go forward.

When it came out, in February 1632, it received a tremendous amount of publicity, and most of the early comments were flattering to the extent of being fulsome. So far, so good. But some of the highest officials of the Church took a very different view, and one of these, of course, was Scheiner, who–according to Castelli–was so furious about the book that he turned white when discussing it. It is impossible to be sure about the part which Scheiner played, but there are reasons for believing that he was one of those who persuaded the censors that they had been duped. If so, then there can be absolutely no excuse for him. He was a competent astronomer, and probably believed in the movement of the Earth even though nothing would induce him to say so.

Opposition from Scheiner and his colleagues was only to be expected, and did not worry Galileo in the least. What did change the entire situation was the attitude of Pope Urban VIII, who, almost overnight, turned from a well-loved friend into a bitter and relentless enemy. It was hard to believe that he could be the same man as the genial, intelligent Cardinal Barberini who had so often dined at the same table with Galileo and joined in scientific and philosophical discussions.

All sorts of theories have been put forward to explain this sudden *volte-face*. The most favoured idea is that Urban identified himself with

Simplicio, the obtuse traditionalist of the *Dialogue* who was generally made to look ridiculous – even though he was usually given the privilege of having the last word. After the lapse of over three centuries it will never be known whether this was in Galileo's mind or not. On the whole it seems rather unlikely, but he was certainly tactless to use the name 'Simplicio' for the champion of the old school of thought, and there was one case in which Simplicio expressed a view which Urban himself had put forward in earlier and happier times. Galileo should have been much more careful.

But if not this, then what? True, the Pope was under political and even military pressure from various directions, but these troubles had nothing whatever to do with Galileo, and to associate them with him was more ridiculous than any of Simplicio's comments about the status of the Earth. One can understand that Urban might have read through the *Dialogue*, expressed disapproval of some parts of it, and given quiet orders that Galileo should be admonished. But his disapproval was anything but mild, and amounted to something akin to blind rage. The one thing which acted as a brake on his emotions was the inward realization that if he persecuted Galileo too violently he would make himself look absurd, both to his contemporaries and to historians of the future. Even in his anger, he was sensible enough to appreciate the implications. Galileo must, he decided, be silenced for good and all, but there was never any serious intention of putting him to the torture or burning him at the stake as Bruno had been burned only just over thirty years before.

One thing was impossible – and that was to undo what the Pope now regarded as serious damage to the old beliefs. Too many people had read the *Dialogue* by now, and when the printer was sent an official order to send all the unsold copies back to Rome he replied that there were none left. At least one of Galileo's main objects had been achieved. It was the second – to persuade the Church that it would be in everybody's interests to stop trying to halt the progress of science – which was doomed to failure.

No doubt Riccardi was having a most uncomfortable time, because, after all, it was he who had authorized the publication of the book, and if Galileo were now to be condemned as a heretic it would be painfully obvious that Riccardi had made a serious error of judgement. The truth of the matter was, of course, that he had not known what to censor, and had been merely left with a nebulous feeling that all was not well. When the Grand Duke Ferdinand of Tuscany protested at the prohibition of a book which 'had been presented by the author himself in

Rome to the hands of the supreme authority . . . edited in every way that pleased the superiors, and, furthermore, examined here according to the order and command of Rome and finally licensed both here and there', it was very difficult to know just what to say. From all accounts Riccardi did his best to pour oil upon what had become very troubled waters, but with a total lack of success.

Things began to move rapidly toward their climax. First, the *Dialogue* was officially re-examined to see whether it really did take the form of Copernican propaganda. Secondly, the Holy Office decided to call Galileo to Rome to be put on trial for heresy. Galileo received the summons in September 1632, and acknowledged it at once. By now, of course, he was well aware that he stood in real personal danger, and that he could no longer count upon anything but open enmity from the Vatican. It must have come as a shattering blow to him.

Obviously he did not have the slightest desire to come back to Rome,

'Eppur si Muove'

Frontispiece from Galileo's Il Saggiatore, *1623*

and he did his best to avoid it. He pleaded illness (with some justification; after all he was now nearly seventy years old, and in mediocre
health); he pointed out that there was an epidemic of plague; he put
forward one excuse after another. But by January 1633 the Holy Office
had lost all patience, and another letter made it clear that the order was
not to be defied. Either Galileo made haste to set out for Rome, or else
he would be taken there in chains and confined in the prisons of the
Inquisition. After a final and completely fruitless appeal to the somewhat
spineless Grand Duke Ferdinand, Galileo resigned himself to the
inevitable. The Pope did at least allow him to stay with Niccolini, the
Tuscan Ambassador, who received him with real kindness when he
arrived in Rome in mid-February, but the outlook was bleak by any
standards. Up to now, Galileo had always had the reassuring thought
that his friend Maffeo Barberini sat in the Vatican. Now, the old
Barberini had been replaced by the relentless Urban VIII.

The only alternative would have been to go to Venice, away from the
immediate clutches of the Inquisition, but this would have been risky
in more ways than one, and on the whole it seemed best to face up to
the ordeal which lay ahead. There can be little doubt that the period
between the autumn of 1632 and the end of the trial, in July 1633, was
the blackest of Galileo's whole life. It was also the only time that he
really lost heart. In one letter he even wrote that the call from the Holy
Office made him 'detest all the time I have consumed in these studies,
by means of which I hoped and aspired to separate myself somewhat
from the trite and popular thinking of scholars'. The mood passed, but
the mental wounds remained.

Galileo had to wait for some weeks before the first interrogation by the
Inquisition, during which time he remained with Niccolini. The fact of
the matter was that the Church authorities needed time to prepare their
case, which rested on very shaky foundations indeed. All that Galileo
had really done was to publish a book which had received the full
approval of the censors; he had not altered a word, and if anyone were
to blame it must logically be Nicolò Riccardi. The only possible loophole
was to be found in the 1616 decree, according to which Galileo was
ordered to abandon the doctrine of Copernicus, and 'not to hold, teach

Opposite: *The Crab Nebula, Messier 1 in Taurus, the result of the supernova
observed in 1054 : photographed with the 200-inch Hale reflector*

Overleaf: *Star Map from John Flamsteed's* Atlas Cœlestis, *1729.*

STELLATUM
ANTIQUUM.

or defend it in any way, orally or in writing'. Yet there were uncertainties here too, because the record of that fateful interview was unsigned; and Galileo protested that the actual instructions given to him by Cardinal Bellarmine had been much less binding.

Still, the 1616 decree was the only way in which a case against Galileo could be trumped up, and at the first interrogation much was made of it. It was then that Galileo made a disastrous miscalculation. When asked why he had not told Riccardi about the order, he said that he had seen no reason to do so — because the *Dialogue* was not a defence of Copernicanism at all. It was, he claimed, designed to show that the arguments in favour of a moving Earth were nebulous and faulty.

A retraction of this sort seems strangely uncharacteristic of the fiery-natured and confident Galileo, but by now the cards were well and truly stacked against him. Up to the last moment he had probably hoped that scientific reasoning would make the Inquisitors think again; now he had to realize that science played no part in the charade. Clear thinking was out of the question, and he was at the mercy of the Holy Office, which was not noted for its tolerance. At the next interrogation he admitted that in some sections the *Dialogue* was biased in favour of Copernicanism, and at a third hearing he was induced to agree that Bellarmine could well have made a definite order, though he had not heard it at the time and so was in no position to tell Riccardi about it.

Then, on 20 June, there was one last interrogation, at which Galileo was examined with regard to his real views about the status and alleged movement of the Earth. It was then, if at all, that torture could have been applied. But the Pope had no intention of going so far; neither had the Inquisitors, and there is no evidence whatsoever that Galileo was even shown the instruments of torture. Yet his spirit was broken, at least for the moment, and by now he was ready to say and do whatever the Holy Office wished.

The Inquisition had done everything it could. A charge, even though obviously contrived, had been made to stick; Galileo had made what could be interpreted as a confession of error, if not of intent; and only the sentence remained. And so on 22 June, 1633, the old scientist was led into the great hall of the convent of Santa Maria Sopra Minerva to hear his fate. The entire Congregation of the Holy Office was there; it

Opposite top: *The Moon seen from Apollo 8 : the dark plain above centre is the Mare Crisium, which is visible from Earth*

Opposite bottom: *Earth from 106,000 miles. North Africa is seen clearly with the Red Sea and Arabia to the right. Apollo 11 mission.*

was a ceremony the like of which had never been seen before, and has never been seen since. One of the world's great pioneers was to be forced to utter a recantation which neither he nor any other intelligent man could possibly regard as either voluntary or sincere.

The judgement of the Holy Office was quite explicit, and Galileo's recantation had been written out. Kneeling, he read, haltingly:

'I, Galileo, son of the late Vincenzio Galileo of Florence, aged seventy, arraigned personally before this tribunal and kneeling before you, most Eminent and Reverend Lord Cardinals, general Inquisitors against heretical depravity in the entire Christian dominion . . . do swear that I have always believed, and do now believe, and with God's help will in the future believe all that is held and taught by the Holy Catholic and Apostolic Church. But whereas, after an injunction which had been lawfully intimated to me by this Holy Office that I must altogether abandon the false opinion that the Sun is the centre of the world and is immovable, and that the Earth is not the centre of the world and moves, and that I must not hold, defend or teach in any way whatsoever, either verbally or in writing, the said false doctrine, and after it had been notified to me that the said doctrine is contrary to Holy Scripture, I wrote and published a book in which I discussed this doctrine which had already been condemned, and presented arguments in its favour without offering any solution, I have been pronounced by the Holy Office to be vehemently suspected of heresy. . . . Therefore, wishing to remove from the minds of your Eminences and all faithful Christians this vehement suspicion justly conceived against me, with sincere heart and unfeigned faith I do abjure, curse and detest the said errors and heresies and generally each and every other error, heresy and sect which is contrary to the Holy Church; and I swear that in future I will never again say or assert, verbally or in writing, anything which might again give grounds for suspicion against me. . . . I swear and promise also to comply with and observe fully all the penalties that have been or may be imposed upon me by this Holy Office . . . I, the said Galileo Galilei, have abjured, sworn, promised and bound myself as above; and in testimony of the truth I have signed the present document of my abjuration with my own hand and recited it word for word in Rome, in the Convent Sopra Minerva, this 22nd day of June 1633.'

There is a legend that as soon as the recantation had been spoken, Galileo murmured, audibly: 'Eppur si muove'—or, in English, 'And yet it' (the Earth) 'does move'. This is a nice story, but the chances of its being true are virtually nil. The circumstances were not right for any such gesture, and Galileo, we may imagine, was in no mood to run any further risks merely for the sake of showing defiance. The story seems to have been first given in a book published in London, in 1757, by an author named Giuseppe Baretti. Most people discounted it, but in 1911 a painting was discovered which showed the scene and contained the three words. The painter concerned, Murillo, lived from 1617 to 1682, and the date of the picture was given as 'circa 1650', so that at least the story was current then. In all probability Galileo did say 'Eppur si muove', but not at the moment when he had been forced to deny all his true convictions in front of the Congregation of the Holy Office. The comment was presumably made later on, and in private rather than public.

Galileo's house (Il Gioiello) in Arcetri, where he died in 1642

Actually, Galileo was to have no more public life. In addition to the act of recantation, he was to be kept in 'formal imprisonment' at the discretion of the Holy Office, and to be prevented from seeing anyone at all except those who were essential to his well-being. He was taken back to the house of the Tuscan Ambassador, and the authorities considered what was to be done with him. On the sixth of July he left Rome for the last time, and after a few months in the palace of the Archbishop of Siena he came to his final home – his own villa at Arcetri, where he could at least see his family. The official order stated that he had to live in solitude, without receiving any visitors, for a period to be decided by the Pope, but even this was better than might have been expected. Galileo could so well have been shut away in the prisons of the Holy Office.

There are reasons for thinking that the Inquisition was glad to see Galileo well away not only from Rome but also from Siena, where he could be in touch with many of the Archbishop's friends. The Pope himself showed no wish to take matters any further; indeed, he seems to have issued a personal directive that Galileo should not be subjected to any physical discomfort, and there is at least a chance that he realized what the verdict of history was likely to be.

The *Dialogue*, needless to say, remained on the Index of Prohibited Books, but this can hardly have caused Galileo any worry, because so many copies of it were in circulation that no Papal edict could make any difference. And once he had reached Arcetri, he was determined to go on with his scientific research, even though so far as the movement of the Earth was concerned he was condemned to permanent silence.

He was over seventy, and it is quite remarkable that in those last years of his life he was able to produce what many people regard as his greatest book of all, the *Discourse on Two New Sciences*. This too is a dialogue, and once more Salviati, Sagredo and Simplicio appear; but the whole theme is different, and the book deals with mechanics and what Galileo called 'local motion', so that it does not concern us here. Neither did it touch upon anything which could upset the Holy Office. Because of the total prohibition upon publishing anything which Galileo wrote, the book had to be 'offered around' in non-Catholic countries, and it was finally published in Holland in 1638, where it was an immediate success.

By then Galileo had lost his sight. He had also lost his elder daughter, Sister Maria Celeste, who had died in the spring of 1634. She was the only one of his children who had been of real comfort and help to him in his ordeal, and her death was a tragic blow from which Galileo never recovered; his other daughter and his son Vincenzio were much less close to him. Loneliness was something which he had to bear, and blindness made matters even worse. For some years the order preventing visitors to the villa at Arcetri was strictly enforced, and anyone who wanted to see Galileo had to go through 'the usual channels'; this applied even to old friends such as Benedetto Castelli, who did manage to see him, but only in the presence of a representative of the Holy Office. Old and infirm though Galileo had become, the Inquisition was taking no chances.

It was only after 1639 that the restrictions were relaxed. A young scholar named Vincenzio Viviani was allowed to go to the villa, and stayed there until Galileo died; in the autumn of 1641 he was joined by Evangelista Torricelli, who became an eminent scientist in his own right and is probably best remembered for his invention of the barometer. Even at this late stage Galileo was still ready to join in scientific discussion, as is shown by his correspondence with Fortunio Liceti, who believed the 'ashy light' on the dark hemisphere of the Moon to be due to a lunar atmosphere rather than to reflected earthlight. But the end was drawing near. In November 1641 Galileo was taken ill; he became weaker and weaker, and on the night of 8 January 1642 he died. Viviani and Torricelli were with him; his death was as serene as his life had been stormy.

Galileo's design for a pendulum, drawn by his pupil Viviani, from Galileo's dictation after he became blind

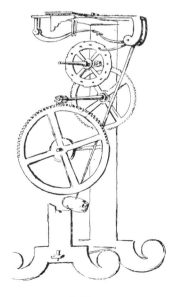

There would be no doubt about the greatness of Galileo even if we considered only his astronomical work and disregarded his experimental mechanics, in which he made even more significant contributions. His life-story, like Tycho's, reads rather like the plot of a sensational novel, but there is no real parallel between the careers of the two men, because Tycho's personal troubles were almost entirely of his own making, whereas Galileo's were not. Of course, Galileo could have avoided persecution if he had refrained from publishing material which was bound, sooner or later, to anger the Church, but in this case his main objective – to spread the truth of Copernicanism – could not have been achieved. On the other hand, he could well have been more tactful. Had he not made so many enemies by his outspokenness, and above all if he had retained the friendship of Urban VIII, things would have worked out differently. Finally, he was badly at fault in trying to pass off the *Dialogue* as an impartial discussion rather than a defence of the theory that the Earth moves round the Sun. Once he had fallen into this trap, nothing could save him from condemnation by the Holy Office.

Not many of the other characters in the drama emerge with real credit. Certainly the Jesuits do not, though the rôle played by Christoph Scheiner has never been made clear and may have been exaggerated (at any rate, let us give him the benefit of what doubts there are). Nicolò Riccardi behaved unintelligently and with some lack of frankness; once he had had second thoughts after originally passing the book for publication he should surely have come out into the open instead of adopting a policy of delay. The authorities at the Holy Office behaved as might have been expected of them. But the key figure was, of course, Pope Urban VIII – Cardinal Maffeo Barberini, who changed sides at the critical moment and so has gone down in history as a treacherous bigot. This may be unfair, but at the very least he mismanaged the entire episode, and the trial and condemnation of Galileo is something which the Catholic Church would be only too glad to forget.

There was one final piece of pettiness. It was proposed to erect a monument over Galileo's tomb, but the Pope expressly forbade anything of the kind.

A change in attitude was slow to come. By 1820 the Vatican was "obliquely accepting that Galileo's theories were correct", but the final verdict was delayed until our own time. In 1982 Pope John II set up a commission to investigate the whole situation. It took the commission ten years to reach a decision; eventually, on 31 October 1992, the Pope cancelled Galileo's conviction for heresy, and declared that he was after all "legitimate son of the Church". It had taken the Church almost 350 years to admit that the Earth is not the centre of the Solar System!

Truth Will Out

Galileo Galilei died in January 1642. Isaac Newton was born in the following December, so that the lives of the two men did not quite overlap, just as Tycho's did not overlap that of Copernicus. It was Newton who put the finishing touches to the revolution, but the main battle was already over. So far as is known, Galileo was the last man ever to be persecuted for spreading the 'false doctrine' that the Earth is a planet moving round the Sun.

Obviously the change from old to new took some time, and until well into the seventeenth century the Ptolemaic theory was still taught at many universities. For instance, there was the case of the Italian astronomer Giovanni Domenico Cassini–often remembered as Jacques Dominique Cassini, because he spent much of his later life in France and became the first director of the Paris Observatory. (He was succeeded by his son, also Jacques.) When Cassini was appointed lecturer in astronomy at the University of Bologna, in 1650, he found that he had to teach the Ptolemaic system, though by then it was also permissible to teach Copernicanism *as an hypothesis*. Not, of course, that Cassini had any doubts about the truth of the matter. He was an expert observer, and it was he who made the first reasonably accurate measurement of the distance between the Earth and the Sun.

The fact that men such as Cassini could talk openly about the movement of the Earth–and in Italy, less than ten years after Galileo's death–is an indication that the scientific influence of the Church was very much on the downgrade. In this respect the *Dialogue* had had precisely the effect which Galileo had wanted. The authorities in Rome were uncertain as to what was to be done. Within a few decades after the famous trial it had become abundantly clear that if they kept firmly to their policy of stifling all teaching of the heliocentric system they would do no more than make themselves look ridiculous; on the other hand they did not want to give the new ideas any official blessing. So they did nothing at all, and let things take their course. (There is, as we have also noted, a more modern parallel which has affected the Catholic Church in particular and the Protestant Church almost as strongly. In 1859 Charles Darwin published his classic book *Origin of Species*, showing that men and apes have common ancestry. Though he never for an instant suggested that a man is descended from a monkey, the idea of any relationship at all caused a religious storm greater than any which had taken place since the time of Galileo. I have referred to the law which prohibited teaching the theory of evolution in parts of America until very recently; even today some branches of the Church are doing their best to gloss over the whole problem. They have no wish to accept Darwinism, but cannot see a way out.)

If I may run ahead of the main story for a moment: it was only in 1734 that the Holy Office at last agreed to allow a monument to be erected to Galileo, and it was ruled that the inscription should contain nothing which might be construed as criticism of the trial or of Rome itself. In 1757 the Church cancelled the decree which prohibited all books relating to the alleged motion of the Earth, but it was not until much later that the works of Copernicus, Kepler and Galileo were taken off the Index. As we have seen, this had to wait until the nineteenth century!

Italian astronomer Giovanni Domenico Cassini, an expert observer who made the first reasonably accurate measurement of the distance between the Earth and the Sun. He became the first Director of the Paris Observatory. (Left)

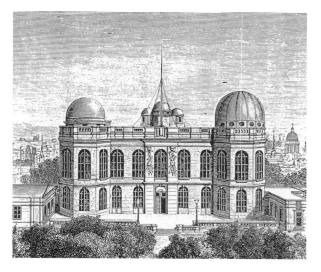

The idea of a moving Earth spread at different rates in different countries. In Russia it gained little headway for many years, and was not widely accepted until the time of the first really great Russian astronomer, Mikhail Vasilevich Lomonosov, in the middle of the eighteenth century. In France it gained some popularity much earlier, and I must here mention the 'vortex' theory of the famous philosopher René Descartes, who was a younger contemporary of Galileo; the whole theory was completely wrong, but it did at least allow for a central Sun. So far as Italy was concerned, the concept of the Earth's motion was admissible as a theory, at least after about 1650, but certainly as nothing more, and there were many dedicated opponents of it.

One of these was a Jesuit named Giovanni Battista Riccioli, who was born at Ferrara in 1598 and lectured successively at the universities of Padua and Bologna. Riccioli had no faith in outlandish theories such as that of Copernicus, but he was by no means averse to looking through telescopes, and in 1651 he produced a reasonably good map of the Moon, based largely upon the observations of his pupil Grimaldi. One thing which occupied his attention was the naming of the craters. On previous maps, various systems had been adopted; but none seemed to be satisfactory, so Riccioli made up his mind to introduce a completely new nomenclature, naming the prominent craters after famous men

The first great Russian astronomer, Mikhail Vasilevich Lomonsov

and women – chiefly scientists. One vast formation, near the apparent centre of the Moon's disk as seen from Earth, was named Ptolemæus; the 56-mile crater in the southern uplands which is the centre of the most widespread system of bright rays or streaks was given to Tycho, for whom Riccioli had a great admiration (his 'system of the world' was essentially the same as Tycho's), and not unnaturally he allotted important formations to himself and Grimaldi. But when it came to the revolutionaries, he was rather less generous. He gave Galileo a very small, obscure crater in the grey plain which was already known as the Oceanus Procellarum, so that, as Riccioli himself said, Galileo was 'flung into the Ocean of Storms' – though Riccioli did relent sufficiently to honour Copernicus with the second of the major ray-craters. We still use Riccioli's nomenclature, though it has been augmented since; and that is why Galileo's name is affixed to an unimportant crater instead of a large and conspicuous one!

In England the situation was quite different, and it was from here that the next developments came. To relate scientific events to English history, it is worth noting that Copernious' *De Revolutionibus* was published during the time when Henry VIII was King of England; Tycho's period at Hven belonged to the reign of Queen Elizabeth I; the telescope was invented, and used by Galileo, while James I sat on the English throne, and at the time of Galileo's trial in Rome the king was the hapless Charles I. (Incidentally, a young astronomer named Gascoigne, a supporter of the moving-Earth idea and also the inventor of an important astronomical measuring device known as a micrometer, was killed at the Battle of Marston Moor, fought between the Cavaliers and the Roundheads in 1642.) Though England had its religious troubles, it was on the whole much less intolerant than many other countries, and men such as Dee and Digges, who made no secret of their pro-Copernican opinions, were not in danger. The heliocentric theory was being taught at Oxford University by 1619, though admittedly still classed as an unproved hypothesis; and an English translation of Galileo's *Dialogue* was widely available by 1661. Things were further helped by the attitude of King Charles II, the 'Merry Monarch' who returned to claim his throne at the Restoration of 1660. No doubt Charles had many faults, but intolerance was not one of them, and he was a sincere lover of science; so too was the famous soldier-statesman Prince Rupert – and both the king and his friends gave strong support to their men of learning. In particular, they backed the formation of the Royal Society, whose early members included Sir Christopher Wren, Robert Hooke, Edmond Halley and, above all, Isaac Newton.

Newton, as we have noted, was born in 1642, eleven months after the death of Galileo, at a time when the status of the Earth was still very much a matter for debate. By the time he died, during the reign of George I in 1727, the whole revolution was over, and no serious astronomer had the slightest doubt that the Earth moves round the Sun instead of *vice versâ*. The revolution would have been complete before 1727 even if Newton had never lived; but because of his immense contributions he must be regarded as part of the main story. There is, too, his famous and perfectly authentic comment that he had seen further than other men because he had been able to 'stand on the shoulders of giants' – that is to say Copernicus, Tycho Brahe, Kepler and Galileo.

Newton's life was far from eventful. He was involved neither in defending his country against invasion (as Copernicus had been), kidnapped as a child and then wounded in a duel (as had happened to Tycho), driven from post to post (as with Kepler) or persecuted and condemned in the manner of Galileo. He took his degree at Cambridge, and spent many years there; his only excursions into politics were a

Isaac Newton who said he had been able to 'stand on the shoulders of giants' that is to say, Copernicus, Kepler and Galileo. Engraving after Kneller

brief spell as a Member of Parliament and his subsequent appointment as Master of the Royal Mint. There was never the slightest hint of official disapproval—indeed, quite the reverse; he received the highest honours of the scientific world, and was knighted by Queen Anne in 1705. Because of this, and because the main controversy came to an end before he published his great book, I do not propose to deal with his personal career in any detail. All I hope to do is to show how he took the earlier research and knitted it together in a form which provided the basis for all later studies.

He came from the little Lincolnshire village of Woolsthorpe, about seven miles south of Grantham. His father died before he was born, and Isaac was so small and sickly that his mother said he 'could have been put into a pint pot'. He went to the village school, where he distinguished himself by his skill at making models but seems to have shown no unusual brilliance; at the age of twelve he went to the King's School in Grantham, where, after a slow start, he learned rapidly. Meanwhile his mother had remarried, but when her second husband also died Isaac was taken away from school and brought home to help in managing the family estate.

The experiment was not a success. By now he had developed a lively interest in mathematics, and farm life did not appeal to him in the least. Mrs. Newton was wise enough to accept the situation. Isaac was sent back to school, and in 1661 he made his way to Cambridge, where he was enrolled as a humble and quite undistinguished student at Trinity College. He matriculated almost at once, and turned his attention to problems of astronomy; he managed to buy a copy of Descartes' book, and mastered it without the slightest difficulty. At this point in his career he was lucky enough to come under the influence of Isaac Barrow, Lucasian professor of mathematics at the University.

Barrow himself had had an eventful life (he was once involved in a shipboard fight against pirates during a voyage off the coast of Turkey!) and he was quick to see the exceptional promise of his new student. Newton was awarded a scholarship, and in January 1665 he took the degree of Bachelor of Arts. In normal times he would probably have taken up an appointment at Cambridge, but conditions in 1665 proved to be far from normal. Plague broke out in London, and claimed thousands of lives. When the disease spread to Cambridge, during the summer, the authorities wisely closed the University and sent the students home. Newton went back to Woolsthorpe, in the same isolation of the Lincolnshire countryside, and stayed there for almost a year; adding the usual University vacations, it is evident that between the summer of 1665 and the summer of 1667 he spent more time at home than at Cambridge. It was during this period that he made the great discoveries which were to lead on to his theory of gravitation.

He was in no doubt about the status of the Earth. It moved round the Sun in a period of one year, and was nothing more nor less than an ordinary member of the Solar System. On the other hand, the laws governing its motion were far from clear; men such as Kepler might have found out 'how' the planets move, but not 'why'. This, of course, brings us on to the story of the falling apple—which, unlike the legend of Galileo and the Leaning Tower of Pisa, is almost certainly true. (It was first related by the French author Voltaire, who said that it had been told to him by Newton's niece.)

Apparently Newton was sitting outdoors one afternoon when he saw an apple fall from the tree on to the ground. As Newton watched, a train of thought started in his mind. There was a definite force which pulled the apple downward; but what was it, and how far did it extend? Gradually, Newton came to the conclusion that the force which pulled on the apple was identical with the force which keeps the Moon in its path. This in turn led on to the idea of 'universal gravitation', according to which every particle of matter attracts every other particle with a force which becomes weaker with increasing distance.

If the apple-tree had been twenty miles high instead of twenty feet, the apple would still have fallen. The same would have been true for a tree two hundred miles in altitude—assuming that one could exist. Next, picture a tree 239,000 miles high. If an apple could be dropped from it, that apple too would presumably fall to the ground. Now, 239,000 miles is the mean distance between the Moon and the Earth. Why, then, does not the Moon fall?

Newton found the answer: the Moon does not fall to the Earth for the simple reason that it is moving. It is not easy to give an everyday analogy, but the old example of a cotton-reel being whirled round on the end of a string is better than nothing, even though it is fundamentally unsound. If you keep the cotton-reel moving, it will go on circling your hand; only if you stop applying a force will it drop. There is nothing to stop the Moon in its path, so that it remains in orbit.

Naturally, this is a gross over-simplification. Kepler had shown that the Moon moves in an ellipse rather than a circle, and in any case we must remember that the two bodies move round the common centre of gravity of the Earth-Moon system, though this 'barycentre' does admittedly lie well inside the Earth's globe. Newton was well aware of this, but he had found the essential clue. The force which pulled the apple was the same as the force acting on the Moon, and, for that matter, the same as the force which keeps the Earth in its path round the Sun.

Now let us return to the cotton-reel. If the string suddenly breaks, the reel will fly off at a tangent, and Newton saw that any moving body will continue its motion in a straight line unless some outside force is acting on it. So far as the Moon is concerned, this force is provided by the gravity of the Earth. In the diagram (Fig 15), the Moon's path is shown (not to scale, for the sake of clarity). But for the Earth, it would move from position M to M1 in, say, one minute. Instead, it moves from M to M2, so that in effect it has 'fallen' from M1 to M2; and it goes on 'falling' all the time, though it comes no closer to the Earth.

Newton calculated that the 'fall' in one minute (that is to say, the distance between M1 and M2) should be 15 feet. Actually, it was only 13. Newton commented that the figures 'agreed pretty nearly', but he was not satisfied, and he felt that something was wrong.

For this sort of calculation, a large body such as the Earth behaves as though all its mass were concentrated at a single point at the centre of the globe. As Newton had to make his observations from the Earth's surface, he had of course to know the distance of the surface from the centre of the globe; in other words he had to know the exact value of the Earth's radius. It has been claimed that the discrepancy of two feet per minute was due to his having used an inaccurate value for the radius, but this seems to be untrue. One link in the chain of mathematical argument was still missing, and was not found until years later.

Another pioneer investigation carried out at Woolsthorpe during the Plague period concerned the nature of light. By passing the Sun's rays through a glass prism, as shown in the diagram (Fig 16), he discovered that light is a mixture of all the colours of the rainbow, and that the different parts of it are bent or refracted unequally—red less than yellow, yellow less than blue and so on. He obtained a full 'spectrum' from red through to violet, but found that when he then passed the light of a single colour through a second prism there was no further splitting-up.

Newton never took this particular research much further, but he was led on to consider the problem of false colour shown by telescopes. This had been a constant source of annoyance to observers ever since the time of Galileo; a brilliant object such as a star would seem to be surrounded by gaudy-hued rings which had no real existence. Newton saw that this was due to the object-glass, which acted rather in the manner of a prism and brought the red part of the light to focus farther away than the blue. The only solution—and a partial one, at that—had been to make the telescopes of tremendous focal length: that is to say, very long. For instance Christiaan Huygens, discoverer of the true nature of Saturn's rings, had a telescope 210 feet in length, so that the tiny object-glass had to be fixed to a mast. Instruments of this kind were incredibly clumsy and awkward to use; it is surprising that they could give any results at all, and Newton cast around for a solution.

He could not see that there was any way of getting round the problem of the unequal refraction of different colours, and so he decided to abandon object-glasses altogether; instead, he would use mirrors. In the 'Newtonian reflector', the light is collected by a curved mirror and sent back on to a smaller, flat mirror placed at an angle of 45 degrees; the flat mirror sends the rays into the side of the tube, where an image is formed and magnified by an eyepiece. Newton's first reflector, a tiny instrument with a one-inch mirror, was presented to the Royal Society in 1671, and it was this which did so much to bring him to the notice of his colleagues.

Actually Newton was not the first to work out the principle of the reflector; an earlier and somewhat different pattern had been proposed by the Scottish mathematician James Gregory in 1663, but Newton's was the first to be made. (When he thought that nothing could be done about the false colour nuisance he was, for once, wrong; compound object-glasses can improve matters, as was demonstrated in the eighteenth century, though false colour can never be completely eliminated.) Newton was no observer, but he had shown the way, and his experiments on the nature of light were just as important as his work on the theory of gravitation.

Isaac Barrow had resigned the Lucasian Chair of Mathematics at Cambridge in 1669, specifically to make way for Newton as his successor, which was a piece of remarkable generosity. Two years later Newton

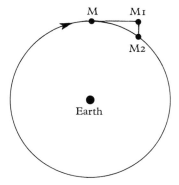

Figure 15: Newton and the motion of the Moon (the 'fall').

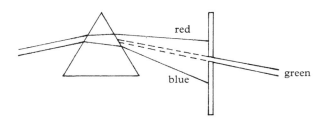

Figure 16: Newton's prism experiment.

became a Fellow of the Royal Society, whose authority in England was supreme. I will merely mention his development of the branch of mathematics which he called the 'method of fluxions' and which we know as the calculus, and add that his first papers to the Royal Society were concerned almost entirely with the nature of light. For the time being he said virtually nothing about his researches into the motion of the Moon. He was sometimes reluctant to publish his results, and one reason for this may have been his uneasy relations with a fellow-member of the Society, Robert Hooke, who was a genius in his own right but who had an unhappy knack of quarrelling with people. Newton disliked criticism, and we must admit that he was frequently over-sensitive.

In 1675, however, he did mention one important discovery in a letter to the celebrated map-maker Nicolas Mercator; and this is worthy of mention here, because it takes us back to Galileo. Writing to Fulgenzio Micanzio in 1637–that is to say during the Arcetri solitude, after the trial in Rome–Galileo had announced that the Moon moves its face in three ways:

'It moves slightly to the right and now to the left; it raises and lowers it, and finally inclines it now toward the right and now toward the left shoulder. All these variations can be seen on the face of the Moon, and what I say is manifest and obvious to the senses from the great and ancient spots that are on the surface of the Moon. Furthermore, add a second marvel: these three different variations have three different periods, for the first changes from day to day and so has its diurnal period, the second changes from month to month and has a monthly period, and the third has an annual period whereby it completes its cycle.'

Galileo went on to draw the completely erroneous conclusion that this behaviour was linked with the tides (by then he had given up his earlier theory, outlined in Day Four of the *Dialogue*). But his observations were valid enough; he had seen what we now call the lunar librations, and Newton explained them (Fig 17).

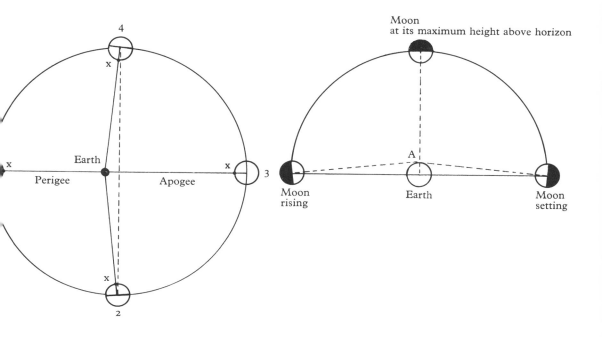

Figure 17: Principle of libration.

Because the Moon's revolution period is 27.3 days, the same as the time taken for the Moon to spin once on its axis, essentially the same hemisphere is always turned toward the Earth. But although the Moon's rate of spin is constant, its orbital velocity is not. It moves quickest when it is at its closest point to the Earth (perigee) and slowest when it is farthest away (apogee). Therefore the position in orbit and

The 150 foot 'aerial telescope' used by Johannes Hevelius; from his Machina Cœlestis, *1673*

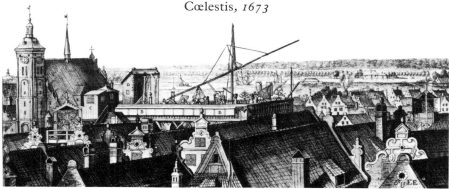

the amount of spin become periodically 'out of step', and the Moon seems to tip very slowly first one way and then the other, so that we can peer for some distance round alternate mean limbs. This is the libration in longitude. There is also a libration in latitude, because the Moon is sometimes north and sometimes south of the ecliptic, and a daily libration due to the fact that we are observing from a point on the Earth's surface almost 4000 miles from the centre of the globe, so that we can take advantage of the Earth's own rotation. In his letter to Mercator, Newton set out the facts and the explanation of them.

Meantime, Robert Hooke had been delving into the problems of gravitation, and had come to conclusions which were much the same as Newton's. He claimed that the attracting force between two bodies must vary 'inversely as the square of the distance from the centres about which they revolve', and this so-called Inverse Square Law is the crux of the whole matter.

It can be explained without using anything but easy arithmetic. Consider the impossible but convenient case of two planets travelling round the Sun, one at a mean distance of 2 million miles and the other at 5 million miles. 2 squared, or 2 x 2, is 4; 5 squared is 25. Then the force of the Sun on the two planets will be as $\frac{1}{4}$ is to $\frac{1}{25}$, and the force on the more distant planet will be only $\frac{4}{25}$ of that on the nearest planet.

If this law be true, it can be shown that each planet will move in an elliptical orbit. Ever since Kepler's day it had been known that the planetary orbits are indeed ellipses, but the full mathematical solution had not been given, and Hooke, brilliant though he was, could not provide it. Neither could the other leaders of the Royal Society, as they found when a series of discussions was held in 1684. Sir Christopher Wren – professor of astronomy at Oxford before turning to architecture – offered to present a forty-shilling book to anyone who could provide the complete proof of the correctness of the inverse square relationship.

Robert Hooke claimed to have the answer, but said that he would

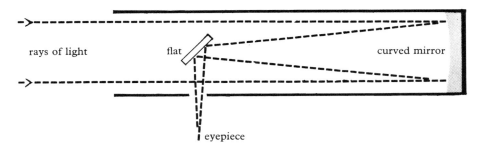

Principle of the Newtonian reflector.

'conceal it for some time, that others trying and failing might know how to value it when he should make it public'. When the months passed by and nothing happened, Edmond Halley went down to Cambridge to consult Newton, who was known to be the most brilliant mathematician in England.

Halley did not mince matters, and asked a straightforward question: 'If the inverse square law be true, what will be the path of a planet?' Newton's reply was equally direct: 'An ellipse.' Halley asked how he could be so sure, and Newton answered: 'Why, I have calculated it.' It must have surprised Halley to find that Newton had had the solution all along without publishing it.

Moreover, it transpired that he had lost his notes. Before leaving Cambridge, Halley had made Newton promise that he would either find the original calculations or else do them again, and Newton was as good as his word. He was held up temporarily by a careless mistake in one of his diagrams, but three months after Halley's visit Newton sent him the proof he wanted, though whether he ever received Wren's forty-shilling book is not on record. Subsequently, Newton gave a short but important paper on the subject to the Royal Society.

Still Halley was not satisfied, and after a great deal of persuasion Newton was induced to write the book which should properly be called the *Principia Mathematica Philosophiæ Naturalis (Mathematical Principles of Natural Philosophy)* but which everyone calls the *Principia*. The whole work took fifteen months to complete, and has been described as the greatest mental effort ever made by one man. Publication arrangements did not go smoothly; this time there was no question of censorship, but further squabbles with Hooke so infuriated Newton that at one stage he threatened to withhold the third part of the book altogether. Hooke was annoyed at Newton's failure to mention his work, and it took all Halley's considerable tact to smooth matters out. Then the Royal Society, which had agreed to finance the publication, ran short of money, mainly because it had recently issued a vast volume by one Francis Willoughby entitled the *History of Fishes*—which had not sold well, and had involved the Society in considerable loss. Eventually Halley paid for the publication out of his own pocket, and saw the *Principia* through the press, for which generosity the world can never be too grateful to him. (Even then Halley had not heard the last of the *History of Fishes*. Later on, when he was a salaried official of the Royal Society, he was owed £50 arrears of pay, and was given fifty copies of the Willoughby book in lieu of hard cash; he was also awarded a further twenty copies for additional back payments. His reactions are not known, and neither have we any idea what he eventually did with his seventy copies!)

It would be out of place here, as well as unwise, for me to make any attempt to summarize the contents of the *Principia*, which finally came out in 1687. I will therefore select only a few of the points which have a direct connection with the main scientific revolution: the laws of motion, the movements of the planets, the precision of the equinoxes, and the cause of the tides.

There are three laws of motion. 1. Every body continues in its state of rest, or of uniform motion in a straight line, unless it is compelled to change that state by forces acting upon it. 2. The change of motion (or 'acceleration') is proportional to the motive force acting, and is made in the direction of the straight line in which that force is impressed. 3. To every action there is always an equal and opposite reaction. I doubt whether these require any further explanation, though I cannot resist pointing out that modern space-ships work according to the principle of reaction laid down in the Third Law; as the gas rushes out of the rocket exhaust it propels the rocket itself in the opposite direction – and this will continue so long as the gas keeps on rushing out, and obviously there is no need to 'push against' a surrounding atmosphere, which is why our probes can function excellently when beyond the top of the Earth's blanket of air. Even the Apollo missions go back ultimately to the work of Newton.

Once these laws had been given, Newton proceeded to lay down the principles of universal gravitation, in the course of which he proved the inverse square law which had so baffled Hooke, Halley and everyone else. In the second book of the *Principia* he discussed the motions of fluids, and laid the foundations of the science which has now developed into mathematical physics, besides giving numbers of brilliant and original experiments, and showing that most of the old ideas about planetary motions were either incomplete or else wrong. Then, in the third book, he investigated the movements of all the bodies in the Solar System, treating them not as mere points but as the great globes they really are. He described the behaviour of the planets; the motions of the Galilean satellites of Jupiter and the five attendants of Saturn which had been discovered by that time (one by Huygens and four by Cassini); the causes of the polar flattening of a planet; the precession of the equinoxes; the tides, methods of finding out the real masses of the Sun and planets, and much else besides.

Precession was explained, correctly, as being due to the pull of the Sun and Moon upon the Earth's equatorial bulge, causing the axis of rotation to 'wobble' slowly and slightly in the manner of a gyroscope which is running down. Of course, precession had been known ever since Greek times, but nobody had previously been able to account for it.

What, then, of the tides? Universal gravitation showed that the Moon's pull is indeed the answer, heaping up the water below it and also on the opposite side of the Earth; when the Moon lies in the same direction as the other tide-raiser, the Sun, we have strong or spring tides, while when the Sun and the Moon are pulling against each other we have the much gentler neap tides. Galileo could never admit that the Moon could act in such a way, so that once again the discovery had to await the genius of Newton (Fig 18).

The *Principia* had an immediate effect, and to all intents and purposes it marked the end of any opposition to the idea of a moving Earth. Henceforth astronomy was taught and studied upon 'Newtonian principles', and the systems of Ptolemy and Tycho became nothing more than historical curiosities. In the years following 1687 Newton's reputation continued to grow, and it has never diminished. True, the *Principia* remained his greatest work, even though we must not forget the later *Optics,* which was published in 1704; but Newton never gave up his scientific researches, and the third edition of the *Principia* was produced when he had reached the advanced age of eighty-four.

He was involved, too, in matters concerning a new catalogue of the fixed stars. At that time there were long discussions about how to solve what was called the 'longitude problem'. During a long sea-voyage it was difficult for sailors to obtain an accurate navigational fix, and this could lead to disastrous results. Finding latitude was easy; all that had to be done was to measure the altitude of the Pole Star, which lies very close to the celestial pole. Longitude-finding was much more troublesome, and British sailors badly needed a reliable method. The only solution seemed to be to use the changing position of the Moon in the

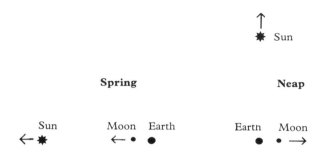

Figure 18: Spring and neap tides.

sky, and this involved knowing the positions of the stars on the celestial sphere, since it was these stars which had to be used as reference-points. Tycho's catalogue was still the best available, but since it had been drawn up before the invention of the telescope it was not accurate enough to be used in precise navigation. Accordingly, Charles II ordered that an observatory should be set up at Greenwich, and the stars re-measured. The Rev. John Flamsteed was put in charge, and was later made Astronomer Royal. He was a first-class observer, but he was not quick, and he was a perfectionist as well, so that years passed by without the promised catalogue being made available. Eventually it was published, and proved to be every whit as good as had been hoped – though, ironically, the invention of an accurate chronometer (that is to say, a timekeeper which could be used at sea) meant that it never had to be used for the purpose originally intended.

Flamsteed, unfortunately, was of a touchy and irritable nature, and he was furious when parts of his catalogue were issued without his authorization. Worse, there was a preface, written by Halley, which was considered as harmful to Flamsteed's reputation. Newton held some of the material, and refused to hand it over; Flamsteed obtained as many copies of the unauthorized publication as he could, and publicly burned them. It was all very regrettable, and none of those concerned in the episode emerged with much credit. I mention it here only to show that scientific disputes of the eighteenth century were very different from those of earlier times, when Bruno had been burned at the stake and Galileo called before the full might of the dreaded Inquisition.

It will, I think, be fitting to end by going back to those ghostly visitors from space, the comets, because it was a comet which provided the final observational proof of all that Newton had said. In 1607 a bright comet had appeared in the sky; this was the time when Galileo was lecturing in Padua, while Kepler was in Prague – and it is on record that Kepler first saw the comet as he stood on one of the bridges overlooking the river Moldau, after he had been watching a firework display. On 15 August 1682 a comet was discovered at Greenwich Observatory, and Edmond Halley had his first view of it during the second week in September; it became a conspicuous naked-eye object, and remained on view for some time before it drew away from the Sun and was lost. Newton, too, looked at it, but was by no means certain whether or not comets could be classed as bona-fide members of the Solar System. In any case their orbits were different from those of the planets, and were either very eccentric ellipses or else open curves; Newton tended to favour the latter. On the other hand, it was at least clear that they were real bodies, and not mere atmospheric phenomena as Galileo had thought.

Later, when the *Principia* had made its impact, Halley determined to make a thorough investigation. Using Flamsteed's observations of the comet of 1682 he worked out its path, and found that it seemed to be in much the same orbit as the comet of 1607. It was also similar to that of a comet observed in 1531. Could they be one and the same? Halley believed so, and therefore it should be possible to use Newton's laws of gravity to work out when and where the comet would come back.

In 1705 Halley wrote to the Royal Society: 'Now many things lead me to believe that the comet of the year 1531, observed by Apian, is the same as that which in the year 1607 was described by Kepler and Longomontanus, and which I saw and observed myself at its return in 1682. All the elements agree, except that there is an inequality in the times of revolution; but this is not so great that it cannot be attributed to physical causes. For example, the motion of Saturn is so disturbed by the other planets, and especially by Jupiter, that its periodic time is uncertain to the extent of several days. How much more liable to such perturbations is a comet which recedes to a distance nearly five times greater than that to Saturn, and a slight increase in whose velocity could change its orbit from an ellipse into a parabola! The identity of these comets is confirmed by the fact that in 1476 a comet was seen, which passed in a retrograde direction between the Earth and the Sun, in nearly the same manner; and although it was not observed astronomically, yet from its period and its path I infer that it was the same comet as that of the years 1531, 1607 and 1682. I may, therefore, with confidence predict its return in the year 1758. If this prediction is fulfilled, there is no reason to doubt that other comets will return.' He added, modestly, that if he were proved right, posterity would not fail to acknowledge that the discovery had been first made by an Englishman.

Halley died in 1742 – the last of all the great characters who had been in any way concerned with the scientific revolution. New calculations made in France modified the date of the comet's return to 1759, and astronomers began to keep a careful watch. Then, on Christmas Night 1758, a Saxon amateur astronomer named Palitzsch had the first glimpse of the returned wanderer, and it duly passed perihelion on 12 March 1759, very much as Halley had expected. The last doubters were silenced, and the triumph of the new astronomy was complete.

Epilogue

I have said that the scientific revolution was not rapid, and if we begin with Aristarchus it took nearly two thousand years. Yet once it was really under way, it speeded up tremendously. The main battle began in 1543, with the publication of Copernicus' *De Revolutionibus,* and the final word was said by Newton in the *Principia* of 1687. If we take these as being the limits, then the whole revolution was spread over only 144 years, which is not very much when we remember that it involved a total revision of all Man's ideas about the universe in which we live. And if we consider only the time between 1543 and the trial of Galileo in Rome, then the spread is restricted to a mere ninety years.

Could it have been achieved even more quickly? Most people claim that the progress of science was delayed by the Catholic Church, which used every trick and every device it could find in order to hold things back. My own view, for what it is worth, is that the Church caused no delay at all, though it certainly tried to do so. True, Copernicus might have published his work a decade earlier than he did, but a few years at that stage made no real difference, since in any case there were no striking developments for a long time afterwards. Neither Kepler nor Galileo held back their work; Kepler had no reason to do so, while Galileo was prepared to go his own way with a disregard for the consequences which amounted almost to foolhardiness. There was a halt between 1616, when he was first warned, and 1624, when he set to work on the *Dialogue* in earnest; but even if there had been no religious opposition at all, the book could scarcely have appeared much earlier than its actual publication date in 1632. With it, Galileo had said all that he was really able to say about the movement of the Earth.

Had the Church been able to continue with its policy of silencing all the opponents of Ptolemy and Aristotle, things would have been very different, but there was no real chance that it could do so. To use a mixed cliché, the Holy Office bolted the door of the Galilean stable only after the horse had escaped. All the same, it may have been fortunate for the spirit of scientific inquiry that Newton was an Englishman rather than an Italian.

In our admiration for those who made the revolution possible, we must not disregard their mistakes. In any major upheaval of thought there are many blind alleys which must be followed before the true path becomes known. The real piece of good luck was that the telescope came upon the scene at the crucial moment; without it there would certainly have been a considerable delay—but this has nothing to do with either the Church or with politics.

Let us, then, sum up the rôles of the leading figures in the story. Copernicus, the Polish canon, made the initial breakthrough by demoting the Earth from its central position and replacing it with the Sun; but his system was wrong in almost every other way, largely because of his faith in perfectly circular orbits and his reliance upon Ptolemaic epicycles and deferents. Tycho Brahe, the impetuous Dane, made his contribution solely in the field of observation; his theories were rooted in the past, but without his painstaking, incredibly accurate measurements of the positions of the stars and the wanderings of Mars there would have been no data upon which his immediate successors could rely. Kepler, unhappy and physically frail, provided the next essential step when he proved that the orbits of the planets were elliptical rather than circular, so making possible the compilation of tables of planetary motion which were far better than anything previously achieved. Galileo, with his tactless but effective propaganda and his magnificent series of telescopic observations, showed the world that the Earth was no more than an ordinary member of the Sun's family— even though he too was wedded to the concept of circular orbits, and his 'conclusive' proof, that of the tides, turned out to be no proof at all. Finally, when the thunder of battle had almost died away, came Isaac Newton, to say the last word.

All these men were indispensable links in the chain, and each, after Copernicus, depended upon the work of his predecessors. They fitted together so perfectly that the pattern might almost have been deliberately worked out. Fate can play some curious tricks.

Undoubtedly the Copernican revolution was the greatest that had taken place in human outlook since the age-old discovery that the Earth is a globe rather than being flat. Since then it has been found that the Sun itself is a totally unimportant member of the Galaxy, and that even the Galaxy is one of millions upon millions of similar systems; but this revolution, vital though it was, cannot be held to be so fundamental as that due to Copernicus, Tycho Brahe, Kepler, Galileo and Newton. So as we look back to those strange, turbulent times, let us pay homage to the men to whom we owe so much. So long as history lasts, our memory of them will never fade.

Epilogue:
From Newton to the Space Age

The controversy was over; the Sun, not the Earth, was the centre of the Solar System. The essential nature of the Moon and planets was recognized; before long the Sun itself was relegated to the status of an ordinary star, and it was clear that the universe was vaster than had been believed. It was in 1672 that Giovanni Cassini made the first reasonable measurement of the distance of the Sun; his value - 86,000,000 miles - was rather too small, but at least it was of the right order. Yet several major problems remained. How far away were the stars? Were there other star-systems as well as our own - and, perhaps the most intriguing of all, how did the universe begin?

It has been said that scientific progress tends to be made in a series of jerks. The sixteenth and seventeenth centuries abounded with great men; then, during the first part of the eighteenth century, came something of a lull. True, telescopes were improved and new observatories were set up, but the next major figure was that of Friedrich Wilhelm Herschel, always known as William Herschel because although he was Hanovarian by birth, he emigrated to England while still a young man and spent the rest of his life there. Trained as a musician, he became organist at the fashionable resort of Bath, but in the 1770s he decided to make a hobby out of astronomy, and before long music became secondary in his life. He built reflecting telescopes which were the best of their time, and it was with one of these, in 1781, that during a "survey of the heavens" he chanced upon a new planet, the world we now call Uranus. It proved to be a giant, moving round the Sun well beyond the orbit of Saturn, the outermost member of the Solar System previously known.

That discovery made Herschel famous. He became King's Astronomer to George III of England and Hanover (not Astronomer Royal; that post was held by Nevil Maskelyne) and he became arguably the best observer of all time. His aim was twofold. He wanted to see how the stars were distributed in space, and he wanted to find out their distances. In the first aim he was as successful as he could reasonably have hoped; he decided that the Galaxy was shaped like a "cloven grindstone", and this is a fair approximation, even though he erred in placing the Sun near the centre of the system. However, his attempt to measure stellar distances failed simply because his measuring equipment was not sensitive enough.

The method he used - that of parallax - was basically the same as that used by a surveyor to determine the distance of some inaccessible object, such as a mountain-peak. The object is observed from the opposite ends of a baseline, and its angular shift noted. If the length of the baseline is known, the distance of the object can be worked out by straightforward trigonometry. Herschel concentrated upon double stars, on the assumption that one member of the pair would be much closer than the other; therefore, observing the relative positions over a six-monthly period would mean using a baseline of 186,000,000 miles, twice the radius of the Earth's orbit. What Herschel did not initially know was that most double stars are binaries, or physically-associated pairs, rather than 'optical doubles', which are mere line of sight effects. It was during the course of his investigations that Herschel discovered the existence of binary systems. It was only in 1838, sixteen years after Herschel's death, that Friedrich Bessel in Germany used the same principle to measure the distance of a dim naked-eye star in the Swan, 61 Cygni, which showed a detectable parallax shift against the background of more remote stars. Bessel found the distance to be about 11 light-years - a light-year being the distance travelled by light in one year: around 5,880,000,000,000 miles.

Other parallax determinations followed, but beyond a few hundred light-years the shifts become swamped in unavoidable errors of observation, and less direct methods have to be used. These were based upon the principle of the spectroscope, which of course goes back to the work of Newton at Woodsthorpe during the Plague years. It was found that the spectra of normal stars are basically similar to those of the Sun, but there are great differences in detail; for example, the spectrum of a bright white star such as Sirius is dominated by hydrogen, while in the cooler orange-red giant Betelgeux we find complex effects due to molecules. Spectroscopic investigations can give clues to the real luminosities of the stars, and from these the distances can be calculated, thought not with precision.

The other major problem of the time concerned the status of the Milky Way Galaxy. In 1781 Charles Messier of France, published a list of nebular objects; some (such as the Pleiades in Taurus) were star-clusters, while others were lumped together as nebula. The nebula were of two distinct types. Some, such as the Sword of Orion, were obviously gaseous, but

235

others, notably the Great Nebula in Andromeda - No. 31 in Messier's catalogue - seemed to be made up of stars. Was it possible that these "starry nebulæ" were independent systems, far beyond ours? Herschel thought it possible, but there could be no proof, because the nebulæ were certainly too far away to show any detectable parallax shifts. In 1845 the Early of Rosse, using his great 72-inch home-made reflector at Birr Castle in Ireland, discovered that many of the "starry nebulæ" were spiral in form, like Catherine-wheels, but this was no real help in clearing the problem up.

It was at this time, too, that another revelation gave extra confirmation to Newtonian theory. Uranus, discovered by Herschel in 1781, was not moving as expected. It persistently wandered away from its predicted path, so that clearly it was being perturbed by some unknown body. Two mathematicians, Urgain Le Verrier in France and John Couch Adams in England, independently came to the conclusion that this unknown body must be a more remote planet, and they set out to find it mathematically. It was a sort of cosmic detective problem - they could see the "victim", Uranus, and they had to find the "culprit". In 1846 a new planet, Neptune, turned up in almost exactly the position given by Le Verrier and Adams - yet another triumph of Newton's mechanics.

Telescopes developed rapidly during the late nineteenth and early twentieth centuries. There were powerful refractors, of which the largest, at the Yerkes Observatory in the United States, had an object-glass 40 inches across; but gradually reflectors took over, particularly when it became possible to make mirrors of glass rather than of metal (as Herschel and Lord Rosse had done). Moreover, a mirror can be supported along its back, whereas a lens has to be supported round its edge, and if too heavy it will tend to distort under its own weight, making it useless. In 1917 came the Mount Wilson reflector, in California, with its 100-inch mirror; it was with this telescope, which for many years was in a class of its own, that Edwin Hubble at last solved the problem of the status of the Galaxy.

Most stars - including the Sun, fortunately for us - shine steadily for year after year, century after century; but there are some which do not. These variable stars brighten and fade over relatively short periods. Among them are the Cepheids, which take their name after the prototype star of the class, Delta Cephei in the north of the sky; these have periods of a few days or weeks, and are as regular as clockwork. In America, Miss Henrietta Leavitt made a careful study of the Cepheid variables in the system known as the Small Cloud of Magellan, which is too far south in the sky to be seen from

European or North American latitudes (Miss Leavitt used photographic plates taken in South America). She found that the Cepheids with longer periods were systematically brighter than those with shorter periods. For all intents and purposes the stars in the Small Cloud could be regarded as being equally distant from us, just as for all purposes it is good enough to say that Victoria Station and Charing Cross are the same distance from New York, and it followed that the longer-period Cepheids really were the more luminous. Miss Leavitt was able to draw up a definite relationship. It followed that once the period of a Cepheid was known - which could be found from simple observation - its luminosity, and hence its distance, could be found.

Cepheids are very powerful stars, thousands of times more luminous than the Sun, so that they can be seen over vast distances. Edwin Hubble used the Mount Wilson 100-inch to detect Cepheids in the Andromeda Spiral and other starry nebulæ. At once it became clear that the Cepheids were much too remote to be members of our Galaxy, so that the spirals were themselves independent systems. A revision to the Cepheid scale made in 1952 by Walter Baade, using the 200-inch reflector which had been set up at Palomar, established that the Andromeda Spiral is over 2,000,000 light-years away from us. Most galaxies are much more remote than this. Some of the systems known as quasars, which appear to be very active galaxies, are at least 13,000,000 light-years away.

There is another almost equally important point. If a source of light is receding, it is slightly reddened; if it is approaching, the light appears slightly more blue than it would do if the source were stationary. This is the well-known Doppler effect. The actual colour-change is slight, but it affects the positions of the dark lines in the spectrum of a star - or of a galaxy, which is a medley of the combined light of many millions of stars. Hubble found that apart from the Andromeda Spiral and other members of what we call the Local Group, all the galaxies are racing away from us - and the further away they are, the faster they are going. In fact, the entire universe is expanding.

Meanwhile, there had been theoretical developments. Einstein's theory of relativity had appeared. Obviously it was to some extent built upon Newton's pioneering; but Newton's principles could go only 'just so far'. New techniques and new instrumentation had been refined and extended and this paved the way for Einstein and a new approach. And just as Newtonian theory had been confirmed by observation, such as the return of Halley's

Comet and the tracking-down of Neptune, so relativity was confirmed observationally. Certain small irregularities in the motion of the planet Mercury were explained by relativistic mathematics; the departure from strict Newtonian theory was very slight, but it was significant. Also, relativity predicted that light passing by a massive body would be slightly bent, and this too was confirmed in 1919, when there was a total eclipse of the Sun. While the Sun itself was hidden by the dark disk of the Moon, stars close to it in the sky were found to be displaced by precisely the amount that Einstein had predicted.

Newton in the seventeenth century, Einstein in the twentieth century ... when will we have another figure of comparable stature? We cannot tell.

The Mount Wilson 100-inch reflector remains the largest in the world for three decades, but in 1948 came the 200-inch reflector set up at Palomar Mountain, also in California. It was master-minded by George Ellery Hale, an American astronomer who not only planned great telescopes but also had the happy knack of persuading friendly millionaires to finance them. The 200-inch reflector was an instant success, but by then there had been another revolution. Visual observation at the telescope eyepiece had been superseded by photography, and by the mid-20th century almost all actual research was carried out photographically.

Then - on 4 October 1947 - came the start of the Space Age. Dreams of reaching other worlds go back a long way (as early as the second century AD a Greek satirist, Lucian of Samosata, wrote a light-hearted story about a flight to the Moon), but until our own time all thoughts of travelling beyond the Earth remained in the realm of science fiction. Obviously, no ordinary flying machines can be used, because they depend upon the presence of atmosphere, and there is very little atmosphere above a height of a few miles. The only device which will function in airless space is the rocket - and here we come straight back to Newton: "Every action has an equal and opposite reaction".

Consider a rocket of the type used in fireworks displays. It consists of a hollow tube filled with gunpowder, with a stick to provide stability. When the powder is lit, hot gas is sent out of the exhaust; this propels the tube in the opposite direction, and so long as the gas continues to stream out so the rocket will go on flying. In fact, it depends upon Newton's principle of reaction. Almost a century ago now a Russian, Konstantin Eduardovich Tsiolkovskii, described the principles of rocket flight, and proposed to construct a true reaction motor, replacing the solid explosive with liquids which could be forced into a combustion chamber, where they would react and produce the gas to make the rocket fly. Tsiolkovskii even suggested mounting rockets on top of each other, so that the upper stage would be given what may be termed a running start into space. Moreover, he realized that the minimum take-off speed must be 7 miles per second, assuming that no extra impetus is to be given. This is the Earth's escape velocity; a body departing at 7 miles per second will not return, because the Earth's gravity will not be powerful enough to draw it back.

Tsiolkovskii was not a practical experimenter, and it was not until 1926 that the first liquid-propellant rocket was fired - not in Russia but by an American, Robert Hutchings Goddard. It was modest enough, rising to a few tens of feet at a maximum speed of 60 miles per hour, but it was the first direct ancestor of the space-craft of today. Subsequently a German group, including Wernher von Braun, established an experimental rocket flying field near Berlin, and made quick progress. In 1937 the group was taken over by the German Government and transferred to Peenemünde, an island in the Baltic, where von Braun and his team developed the V.2 weapons used to bombard London during the final stages of the war. After the German collapse, the rocket team went to America and continued with their work, this time along more peaceful lines. By 1949 some step-rockets had risen to well over 200 miles, and in 1955 the Pentagon announced that an artificial satellite would be launched in the immediate future.

Yet again we return to Newton's principles. If a projectile is given what is termed orbital velocity - 5 miles per second - it can enter a path round the Earth, and will remain in orbit. It will obey Kepler's Laws, provided that it is moving above the top of the atmosphere and is not being braked by resistance against air-particles. It seemed feasible to make the attempt, but in fact the first success came from what was then the Soviet Union. On 4 October 1957 the Russians sent up Sputnik 1, the first of all man-made moons. It was football-sized, and carried little apart from a radio transmitter, but it was immensely significant.

There followed what was unashamedly a "space race". Larger and more ambitious Russian satellites went into orbit; the American team, headed by von Braun, managed to send up a satellite in 1958, and indeed it was this vehicle, Explorer 1, which made the initial scientific discovery of the Space Age by detecting zones of charged radiation encircling the Earth, now known as the Van Allen zones in honour of James Van Allen, who designed the satellite's instrumentation. Subsequent progress was remarkably rapid.

Satellites began to be used for mundane purposes such as communication and weather forecasting as well as research into cosmic rays and other phenomena which cannot be studied from ground level because of the screening layers in the upper air. The first manned flight was achieved in 1961, when Yuri Gagarin of the Soviet Union completed a full orbit of the Earth in his cramped module Vostok 1, and by the mid-1960s there had been many flights, often with space-craft carrying several crew members. When President Kennedy announced that America planned to put a man on the Moon before 1970, people no longer laughed.

Obviously the Russians did not intend to be outmatched, but their manned lunar programme foundered simply because their rocket launchers were not reliable enough. So the honour of priority went to America; in 1968 came the first circum-lunar flight, and in July 1969 first Neil Armstrong, then Buzz Aldrin stepped out on to the bleak rocks of the lunar Sea of Tranquillity. Armstrong's "one small step" will never be forgotten; the gap between our world and another had at last been bridged. Further Apollo missions followed, and by the end of the programme, in December 1972, our knowledge of the Moon had been increased beyond all recognition. The value of manned exploration has often been questioned, but it is surely an essential part of space research; men can still do what robots cannot!

Pioneer space-stations were established during the 1970s; the United States Skylab, launched in 1973, remained in orbit for some years, and was manned by three successive crews who used scientific equipment of all kinds, including telescopes. In this field, the USSR was even more successful, and their orbital stations were able to provide homes for their cosmonauts for extended periods. Indeed, by the end of 1988 two crew members, V. Titov and M. Manrov, were able to complete more than a year in orbit - and to come down unharmed.

Gradually it became clear that if space-stations and bases on the Moon were to be set up, it was essential to produce some kind of "ferry" which could be used over and over again. This was the concept of the Space Shuttle, which may be said to take off like a rocket, fly like an aircraft, and land like a glider. Unfortunately, the Shuttle took much longer to develop than had been expected, and there were seemingly endless problems - culminating in a major tragedy in 1986, when the *Challenger* Shuttle exploded soon after take-off and killed all the members of its crew. The Russian equivalent, *Buran*, was also slow to develop, and the abrupt break-up of the Soviet Union has thrown iteffuture into the melting-pot. Candidly, it cannot be said that space-shuttles are yet reliable or really efficient, and it is for this reason that the progress of exploration has apparently slowed down in recent years.

On the other hand, scientific research with unmanned vehicles has been strikingly successful. Many of the most important radiations from beyond the Earth are blocked before they can reach ground level; for example we cannot study X-rays, gamma-rays, or many of the ultra-violet and infra-red radiations. It is here that space research methods really come into their own. X-ray astronomy could not begin until 1962, when rockets became sufficiently efficient to carry equipment above the shielding layers of atmosphere; special satellites were launched to study other regions of the total range of wavelengths, and the results have been spectacular. For instance, the infra-red vehicle IRAS, which orbited the Earth and sent back data during most of the year 1983, surveyed the entire sky in infra-red, and established that some stars are associated with cool material which may well be planet-forming. Evidence is growing that planetary systems are likely to be common in the Galaxy; after all, why should our own ordinary, undistinguished Sun be unique in being attended by a system of planets? And in this case, why should not life also be commonplace?

In the early days of space research the Russians took the lead, and they were also the first to launch probes to the Moon and planets. In 1959 their vehicle Lunik 3 went right round the Moon and sent back the first pictures of the far side of the Moon which can never be seen from Earth because it is always turned away from us (as Galileo had realized so long before). Not surprisingly, the far regions proved to be just as mountainous, just as crater-scarred and just as barren as the side we have always known. Subsequently the United States Orbiters were able to map the entire surface, and by now we know the details of the Moon even better than some of the more inaccessible places on Earth. For a while, after the end of the Apollo programme, the Moon was neglected, but in 1994 a new probe, Clementine, was sent up to provide new maps of the lunar poles, which had not been covered as well as the rest of the surface.

Rockets can reach the Moon in a matter of a day or two. The planets are much further away - even Venus is always at least a hundred times as remote as the Moon - and moreover it is not possible to go by the shortest route; this would involve using propellant throughout the journey, which is out of the question. What has to be done is to put the probe into what is termed a transfer orbit, using the Sun's gravity and "coasting" unpowered for most of the way. To reach an inner planet, the probe must be slowed down relative to the Earth, so that it will swing inward and meets its target planet at a pre-computed point; to reach a planet further away from the Sun than we are, the probe must be speeded up, so that the transfer orbit will take it outward. termed a transfer orbit, using the Sun's gravity and "coasting" unpowered for most of the way. To reach an inner planet, the probe must be slowed down relative to the Earth, so that it will swing inward and meets its target planet at a pre-computed point; to reach a planet further away from the Sun than we are, the probe must be speeded up, so that the transfer orbit will take it outward. Inevitably, a journey will take months - or, in the case of the distant planets, years.

As usual the Russians took the initiative, and their first attempt was made in 1961, when they dispatched a probe toward Venus. However, at that stage, the Soviets were having great problems with long-range communication, and the space-craft was lost after a few weeks. Success came in the following year with America's Mariner 2, launched from Cape Canaveral in Florida; it passed within 22,000 miles of Venus and sent back information showing that, beautiful though it looks in the sky, the "Planet of Love" is not a friendly place. It is intolerably hot, the dense atmosphere consists mainly of carbon dioxide, and the shining clouds are rich in sulphuric acid. It was also confirmed that Venus spins very slowly. The orbital period is almost 225 days, but the axial rotation period is 243 days, and the spin is from east to west, in the sense opposite to that of the Earth and most of the other planets. The reason for this curious state of affairs is unknown. It has been suggested that in its early history Venus was struck by a massive body and literally 'tipped over'; this does not seem very plausible, but it is difficult to think of anything better.

Subsequently the Russians were able to make controlled landings on the surface of Venus, and to obtain direct pictures showing a red, rock-strewn landscape. Much more recently the US Megallan probe has orbited the planet and mapped the entire surface by means of radar. There are highlands, valleys and volcanoes which are probably active; lava-flows are widespread. Certainly there can be no life there, and it seems most unlikely that there will be any manned missions to Venus at least in the foreseeable future. Mercury, the other inner planet, was by-passed in 1974 by another probe, Mariner 10, but - as expected - proved to be cratered and sterile, though rather surprisingly it did turn out to have a detectable magnetic field.

Mars was first surveyed in 1965 by Mariner 4, and proved to be a cratered world more like the Moon than like the Earth. Later missions revealed huge volcanoes, one of which, now known as Olympus Mons, rises to three times the height of our Everest and is crowned by a 40-mile caldera.

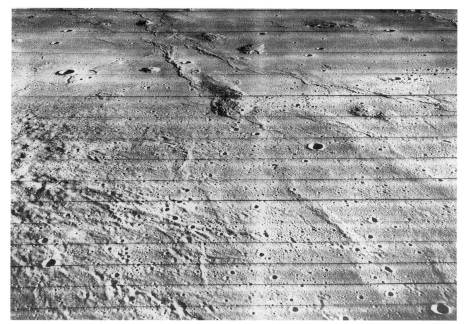

Lunar scenery. An oblique view of the Ocean of Storms (Oceanus Procellarum) on the Moon, photographed from the Orbiter 2 probe in 1966.

Man on the Moon. Colonel Buzz Aldrin stands on the Moon: July 1969. The lunar module of Apollo 11, Eagle, *is in the background. The photograph was taken by Neil Armstrong.*

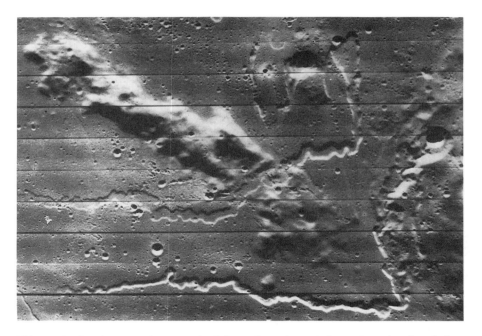

*Old riverbeds of Mars, photographed from the Viking Orbiter. It is hard to believe that **any substance except** water can have been responsible.*

The surface of Mars: the first photograph obtained from the Martian surface, by Viking 1 in July 1976. It shows a rock-strewn landscape; part of the space-craft can also be seen.

The famous dark markings proved to be nothing more than areas where the reddish, dusty material of the 'deserts' has been blown away to reveal the darker surface beneath. They were certainly not old seabeds filled with vegetation, as had been widely believed.

Controlled landings were made in 1976 by the two Viking probes. Material was scooped up from the surface, drawn inside the space-craft, and subjected to chemical analysis. Many people had expected signs of low-type life, but there was no definite sign of any organic activity, and it is now generally believed that at the present time Mars is sterile. On the other hand, there are features which seem certainly to be old riverbeds, so that there must once have been running water; it follows that the Martian atmosphere must then have been much denser than it is now, when the surface pressure is below 10 millibars everywhere. It is not impossible that life did gain a foothold, dying out when conditions became too hostile. We will know only when we manage to obtain samples from the planet itself; this may well be achieved within the next few years.

It is worth noting that the Russians have had no real success with their Mars probes, and neither have they attempted to send missions to the outer part of the Solar System, so that here we owe virtually all our knowledge to American space-craft - in particular Voyagers 1 and 2, launched in 1977.

By a fortunate chance, the four giant planets, Jupiter, Saturn, Uranus and Neptune, were arranged in a long curve during the 1970s, and this meant that it was possible to send a probe on a multi-planet trip. Voyager 1 by-passed Jupiter in 1979, and used the powerful Jovian gravitational pull to send it on to a rendezvous with Saturn in 1980. Voyager 2 was even more ambitious; it surveyed Jupiter in 1979, Saturn in 1981, Uranus in 1986 and finally Neptune in 1989, obtaining magnificent pictures of all four plus a tremendous amount of miscellaneous information.

All the giants provided their quota of surprises. Jupiter is much the most considerable member of the Sun's family; it is over 300 times as massive as the Earth, but it is a very different sort of world. Apparently there is a hot silicate core, surrounded by layers of liquid hydrogen which are in turn overlaid by the dense, hydrogen-rich atmosphere which we can see and which is always changing; Jupiter is in constant turmoil. There are cloud belts, bright zones, and many spots, one of which - the Great Red Spot - is a huge oval with a maximum surface area greater than that of the Earth. It is a whirling storm, and though it may not be permanent it is certainly long-lived; it was definitely recorded by Robert Hooke, so that observations of it date back to the time of Newton. Jupiter was found to have a very powerful

243

magnetic field, and it is surrounded by zones of radiation which would be instantly lethal to any astronaut foolish enough to venture inside them. Jupiter is a world to be viewed from a respectful distance!

Voyagers 1 and 2 sent back close-range pictures of all four Galilean satellites, which proved to be of immense interest. Ganymede and Callisto are icy and cratered; Europa icy and smooth, while Io is orange-red and violently volcanic. The surface is sulphur-coated, and eruptions are going on all the time. Finally, Jupiter was found to have a thin, narrow ring, and the total number of known satellites is now sixteen, though apart from the Galileans all are very small.

Saturn was the next target. Here the main surprise came from the ring system, which turned out to be much more complicated than had been expected; there are many hundreds of ringlets and narrow divisions. The globe itself is of the same type as that of Jupiter, though it is less active. Of the satellites, much the most interesting is Titan, which had been discovered by Christiaan Huygens as long ago as 1655. It is larger than our Moon, and almost as large as Mercury, though less massive; Voyager confirmed that it has a dense atmosphere, made up largely of nitrogen together with a good deal of methane. The low surface temperature almost certainly rules out any forms of life there, but we still do not know what the surface is like; there may even be a wide ocean - not of water, but of some chemical substance such as methane or ethane. We may know, in 2004, when a space-craft, appropriately named in honour of Huygens, is scheduled to make a controlled landing.

Uranus, by-passed in 1986, is a much blander world, containing more icy materials and less hydrogen than Jupiter or Saturn; its diameter is just over 30,000 miles. It has a thin but extensive ring system, and fifteen satellites, though all are fairly small (even Titania, the largest, is less than 1000 miles across). One strange feature of Uranus is that the axis is tilted to the perpendicular by more than a right angle, so that there are times when one pole faces the Sun and has a period of daylight lasting for 21 Earth years. Like all the giants, Uranus is a quick spinner, with a rotation period of 17.2 hours. The orbital period is 84 Earth years.

Voyager 2's main mission was completed by the pass of Neptune, in 1989; by then it had been in space for more than twelve years and had covered over four and a half thousand million miles - yet it was still functioning perfectly. Neptune is a more dynamic world than Uranus, with a bright blue surface and one huge feature known as the Great Dark Spot; its ring system is obscure, and it does not share the curious axial tilt of Uranus. It has eight satellites, but only one, Triton, is large. Voyager 2 obtained close-range pictures showing that the visible pole was coated with nitrogen snow; active nitrogen geysers were also detected.

Voyager 2, like Voyager 1, will never return. It is on its way out of the Solar System, and although we should be able to maintain contact with it well into the 21st century, we are bound to lose it before very long. Millions or even thousands of millions of years from now it may still be wandering among the stars, unseen, unheard and untrackable. It carries plaques and recording which should indicate its place of origin in the event of its being found by an alien civilization, but the chances of t his do not seem to be very high!

This leaves Pluto as the only planet which has not yet been surveyed by a space-craft. It was discovered in 1930 by Clyde Tombaugh from the Lowell Observatory in Arizona; it is smaller than the Moon, and has a companion, Charon, half its size. It has an orbital period of 248 years, and its eccentric path can bring it closer-in than Neptune; perihelion was passed in 1989, and not until 1999 will it again become "the outermost planet". In fact, it seems to be too small to be worthy of true planetary status, and it may be better classed as a "planetesimal", a piece of material left over when the main Solar System was probably formed four and a half thousand million years ago. This could also be the real nature of Triton, which was probably an independent body captured by Neptune long ago; and in very recent years some smaller bodies have been found in this remote part of the Solar System. These may well come from a cloud of planetesimals not far beyond the orbits of Neptune and Pluto.

The minor planets or asteroids are small bodies, most of which move round the Sun between the orbits of Mars and Jupiter; the first and largest, Ceres, was discovered in 1801, but is less than 600 miles across. Most of the rest are true midgets. Two of them, Gaspra and Ida, have been imaged from the latest Jupiter space-craft, named in honour of Galileo and which was due to survey Jupiter in 1995. All the asteroids are cratered, so far as we can infer, and represent débris which was never collected into a larger planet.

Comets have also been studied from space-craft. Halley's Comet returned on schedule in 1986, and was greeted with a veritable armada; two Russian probes, two Japanese and one European. The European space-craft, Giotto, penetrated the comet's head and sent back close-range pictures of a dark nucleus shaped rather like a peanut, with a long diameter of no more

Jupiter, photographed with the Hubble Space Telescope; the Great Red Spot is seen to the lower right, and also on the right the Galilean satellite Europa is disappearing behind the lunar disk.

Neptune, imaged by Voyager 2 in 1989. Voyager 2 is the only probe to have encountered Neptune; this picture was taken from 12,000,000 miles. During the 1.6 hours between the left and right images, the Great Dark Spot has completed a little less than one revolution of Neptune.

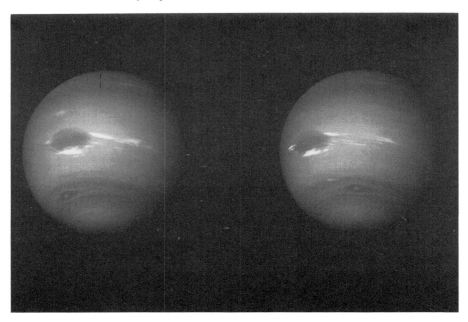

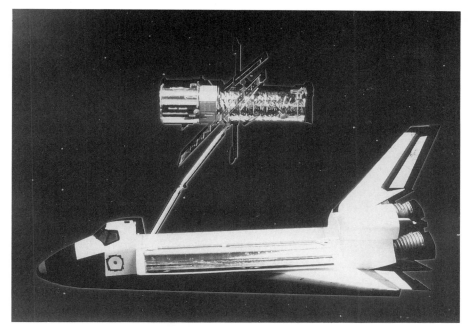

The Hubble Space Telescope, together with the Space Shuttle which launched it into orbit. The height above Earth is over 300 miles.

Remote galaxies, photographed from the Hubble telescope.

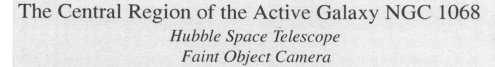

The Central Region of the Active Galaxy NGC 1068
Hubble Space Telescope
Faint Object Camera

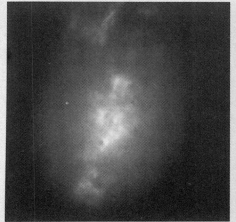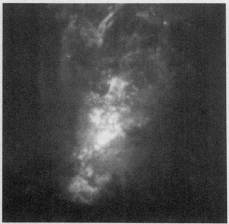

Pre-COSTAR   With COSTAR

than 9.3 miles. After the encounter Giotto was sent on to survey a much smaller comet, Grigg-Skjellerup, though during the Halley pass the camera had been put out of action and no further images could be obtained. Halley's Comet was still under observation in 1994, by which time it had moved out beyond the orbit of Uranus; it may be possible to keep it in view all the way round its orbit.

This should be achieved because of the latest revolution - the onset of the Electronic Age. Just as photography superseded visual observation, so photography is itself now being superseded by electronic devices, which are far more sensitive. This means that large telescopes built years ago, such as the Palomar 200-inch reflector, are more effective now than they were originally. But in any case, the latest telescopes are almost amazingly sophisticated.

For example, the NTT or New Technology Telescope at La Silla, in the lofty Atacama Desert of Northern Chile, has a 141-inch mirror which can be adjusted by computers so that it maintains perfect optical shape no matter at what angle the telescope is pointing. Conditions at La Silla are extremely good; the modern tendency is to site telescopes on the tops of high peaks, so as to rise above the densest part of the atmosphere. Mauna Kea in Hawaii is an extinct volcano with a height of 14,000 feet; a major observatory has been established there, including the new Keck Telescope, which has a 388-inch mirror made up of separate segments which have been fitted together to make the correct curve, A second Keck is under construction. When the two are operating together, they could in theory distinguish the headlights of a car, separately, from a distance of 16,000 miles. We have come a long way since Isaac Newton made his tiny 1-inch reflector little over three centuries ago.

Both Britain's largest telescopes have been set up on the summit of Los Muchachos, yet another extinct volcano but this time in the island of La Palma in the Canaries. The 100-inch is known as the INT or Isaac Newton Telescope; the 165-inch has been named after William Herschel. Both have been outstandingly successful. But, of course, the supreme telescopic achievement - so far - has been the Hubble Space Telescope, named in honour of Edwin Hubble, who was the first to prove that the "starry nebulæ" are external systems rather than being minor features of our own Galaxy. The HST has a 94-inch mirror. It was launched by the Shuttle, and put into an orbit over 300 miles above the Earth, where seeing conditions are perfect all the time. Originally the mirror was faulty, but in 1993 a servicing mission put matters right. Astronauts went to the telescope, captured it, drew it into the Shuttle bay, repaired it and re-launched it, so that the telescope is as good as had been initially expected; though it has "only" a 94-inch mirror it can far outmatch any Earth-based instrument.

Solar System studies have not been neglected; for example, the Hubble space telescope can monitor the volcanoes of Io, and produce images of the planets which rival those of the space-craft. But its main value lies in studying images far away in space. To give even a brief account of its achievements would take many pages, but a few examples may show how powerful it is. In the constellation of Orion we find the Great Nebula, a mass of dust and gas 1500 light-years away - so that we now see it as it used to be 1500 years ago; ever since then its light, moving at 186,000 miles per second, has been racing towards us. In the swirling nebulosity the Hubble space telescope has detected stars which are associated with disks of dust which may indicate planet-forming material. Another Hubble picture shows quickly-moving gas inside the giant galaxy known as Messier 87, which is over 40,000,000 light-years away. The movement of the gas indicates that it is being pulled upon by a very small, incredibly massive object which is presumed to be a black hole - that is to say a region round an old, collapsed star, where the gravitational pull is so strong that not even light can escape from it. Black holes have long been theoretical possibilities, but only now have we the first reliable evidence that they really exist.

Moreover, the telescope can reach out to great depths in the universe, where we find not only conventional galaxies but also the intriguing quasars, which are very small and amazingly powerful; they seem to be the nuclei of very active galaxies. All this brings us on to one of the most fundamental problems of all: how did the universe begin?

In the time of the "great revolution" nobody could seriously speculate about the origin of the universe, but today the whole situation is different, mainly because we can look back into the past. Observe a galaxy or quasar which is, say, 10,000 million light-years away, and we are seeing it as it used to be when the universe was comparatively young. We have seen that the further away a galaxy is, the faster it is receding. If the rule of "the further, the faster" holds good, we will eventually come to a distance at which a system will be receding at the full speed of light. We will then be unable to see it, and we will have reached the boundary of the observable universe, though not necessarily of the universe itself. So far as we can tell, the limiting distance lies somewhere between 15,000 million and 20,000 million light-years, probably closer to the lower value. It follows that the universe as we know it cannot be more than 20,000 million years old.

But we are in a quandary. If the universe began at a set moment in time, with a "Big Bang" (to use the popular term), then what happened before that? We can only say that of time also began at this moment., there was no "before". The only alternative is to assume that the universe has always existed, in which case we have to try to visualize a period of time which stretches back infinitely into the past.

In 1949 a group of astronomers at Cambridge University put forward the "steady-state" theory, which rejected the idea of a Big Bang; they assumed that there was no beginning, and that as old galaxies die new systems are created from material which appears spontaneously out of nothingness. The theory was popular for a while, but current evidence is against it. By observing very remote systems we can in effect look back into the past, and it seems that in these regions the galaxies are not distributed in the same way as they are closer to us; this means that the universe is not in a steady state. Moreover, we have detected weak radiation coming toward us from all directions all the time, and this seems to be the last detectable manifestation of the Big Bang. It indicates a general temperature of about 3 degrees above absolute zero - absolute zero being the lowest temperature that there can possibly be.

One problem was that the background radiation seemed to be absolutely uniform. Presumably then, the Big Bang had also been "smooth"; but how could a non-smooth universe, with galaxies and stars, evolve from such a beginning? What was needed was the detection of slight irregularities in the background radiation, showing that after all the early universe was not absolutely smooth. Success came in April 1992 by using COBE, the Cosmic Background Explorer satellite, which measured tiny but definite differences in the density of the remote material over different regions. It was in fact studying the universe as it had been only half a million years after the Big Bang; there were "ripples", or rarefied wisps of material, which seemed to qualify as the largest and oldest structures known in the universe. Theory had been verified by observation, but this would not have been possible without using space research methods.

What of the future? Even larger and more powerful telescopes are being planned; new satellites and space-probes will be sent up; a fully-fledged orbital station may be in existence within the next ten years, and it is no longer in the least fanciful to discuss a permanent base on the Moon. We can even start thinking about manned missions to Mars.

Yet can we be sure that we are not making some fundamental errors, and that several of our cherished theories may have to be drastically modified or even abandoned? The answer can only be: No. There are suggestions that our distance-measures beyond our local part of the universe may be wrong, and though this is at prevent very much a minority view it has the backing of several leading astronomers. So there may be another major revolution in outlook before long; we can only wait and see.

At least we have more to guide us than would have seemed possible before modern technology came to the door. Yet we always have to look back to the great figures of the past, who did so much to show us the way; without them, we would still be groping blindly in the dark.

CHRONOLOGY OF THE SCIENTIFIC REVOLUTION

General	Copernicus	Tycho Brahe
624 BC Thales born.		
350 BC Aristotle fl.		
280 BC Aristarchus fl.		
230 BC Eratosthenes fl.		
140 Ptolemy writes the *Almagest*.		
1270 Publication of *Alphonsine Tables*.		
	1473 Copernicus born.	
	1491 Copernicus at Cracow.	
	1496 Copernicus at Bologna.	
	1497 Copernicus made Canon of Warmia.	
	1500 Copernicus in Rome.	
	1501 Copernicus in Padua.	
	1503 Copernicus returns to Lidzbark.	
	1507 *The Commentariolus*.	
	1512 Death of Łukasz Watzenrode. Copernicus at Frombork.	
	1516–19 Copernicus at Olsztyn.	
	1520–1 War against the Teutonic Knights.	
	1533 (?) *De Revolutionibus* completed.	
	1539 Rhæticus arrives at Frombork.	
	1543 *De Revolutionibus* published. Death of Copernicus.	
		1546 Tycho born.
		1559 Tycho at Copenhagen.
		1563 First observation by Tycho.
		1571 Tycho in Denmark.
		1572 Supernova in Cassiopeia.
		1573 Tycho's *De Stella Nova*.
		1576 Tycho at Hven.
		1577 Tycho's observations of the comet.
		1596 Tycho leaves Hven.
1600 Bruno burned.		
		1601 Death of Tycho.
1757 Papal Decree cancelled.		
1835 Books by Copernicus and Galileo no longer on the *Index*.		

CHRONOLOGY OF THE SCIENTIFIC REVOLUTION

Kepler	Galileo	Newton
	1564 Galileo born.	
1571 Kepler born.		
	1581 Galileo at Pisa as a student.	
1589 Kepler at Tübingen.	1589 Galileo at Pisa as a professor.	
1594 Kepler at Graz.		
1596 Kepler's *Mysterium*.		
1600 Kepler joins Tycho in Prague.		
1604 Kepler observes a supernova.		
1606 Kepler's *De Stelia Nova*.	1609 Galileo makes his telescope.	
1609 Kepler's *Astronomia Nova*.	1610 Galileo discovers the satellites of Jupiter.	
1610 Kepler's *Conversation*.	Galileo's *Sidereus Nuncius*.	
	Galileo at Florence.	
	1611 Galileo goes to Rome.	
1612 Kepler at Linz.	1612 Attack on Galileo by Lorini.	
	1614 More serious attack by Caccini.	
1615–21 Kepler's mother tried as a witch.		
	1616 The Decree in Rome.	
	1618 Controversy with Grassi about comets.	
1619 Kepler's *Harmonice Mundi*.		
1620 Final volume of the *Epitome*.	1623 Galileo's *Il Saggiatore*.	
	Barberini made Pope.	
	1624–9 *The Dialogue* written.	
1626 Kepler at Ulm.		
1628 Publication of *Rudolphine Tables*.	1630 Galileo takes the *Dialogue* to Rome.	
1630 Death of Kepler.	1632 *Dialogue* published.	
	Galileo called to Rome.	
	1633 Galileo tried and sentenced.	
1634 Publication of the *Somnium*.	Galileo at Arcetri.	
	1638 Galileo's *Discorsi*.	
	1639 Viviani at Arcetri, joined in 1641 by Torricelli.	1642 Newton born.
	1642 Death of Galileo.	1661 Newton at Cambridge.
		1665–6 Newton at Woolsthorpe.
	1992 Galileo's conviction for heresy cancelled by Pope John Paul II.	1669 Newton made Lucasian Professor.
		1671 Newton presents his reflector to the Royal Society.
		1685–6 Newton writes the *Principia*.
		1687 *Principia* published.
		1704 Newton's *Optics*.
		1705 Newton knighted by Queen Anne.
		1726 Third edition of the *Principia*.
		1727 Death of Newton.

INDEX